VECTORS

VECTORS
ESSENTIAL DATA

Peter Gacesa and Dipak P. Ramji

School of Molecular and Medical Biosciences, University of Wales, Cardiff, UK

JOHN WILEY & SONS

Chichester · New York · Brisbane · Toronto · Singapore

Published in association with BIOS Scientific Publishers Limited

©BIOS Scientific Publishers Limited, 1994. Published by John Wiley & Sons Ltd, Baffins Lane, Chichester, West Sussex PO19 1UD, UK, in association with BIOS Scientific Publishers Ltd, St Thomas House, Becket Street, Oxford OX1 1SJ, UK.

British Library Cataloguing in Publication Data
A catalogue record for this book is available from the British Library.

ISBN 0 471 94841 1

Typeset by Marksbury Typesetting Ltd, Bath, UK
Printed and bound in UK by H. Charlesworth & Co. Ltd, Huddersfield, UK

CONTENTS

Contents

Contents

Index **154**

If you are interested in a disk containing the vector maps in Chapter 3, please contact the publisher, at the following address, for details.
Lisa Tickner, John Wiley & Sons Ltd, Baffins Lane, Chichester, Sussex PO19 1UD, UK. Tel. (01243) 779777. Fax (01243) 775878.

Contents

ABBREVIATIONS

(NB See Chapter 4, Table 4 for key to bacterial genotypes.)

5s	see rrnBT1T2
ACMNPV	*Autographa californica* nuclear polyhedrosis virus
AD	adenovirus
Ap	ampicillin resistance
ars	autonomous replicating sequence
ATG	start codon for protein biosynthesis
bp	base pairs
BW	blue/white selection
CaMV	cauliflower mosaic virus
cat	chloramphenicol acetyltransferase gene
cen	centromere sequence
CITE	CAP-independent translation enhancer
Cm	chloramphenicol resistance
CMV	cytomegalovirus
ColE1	colicin E1 immunity

kb	kilobases
Km	kanamycin resistance
lac	*E. coli lac* operon
λcro	Lambda phage *cro* repressor protein
λP$_R$	Lambda phage promoter controlling synthesis of the R1 transcript
LTR	long terminal repeat
luc	luciferase gene
mam.	mammalian
MCS	multiple cloning site
MMLV	Moloney murine leukemia virus
MMTV	murine mammary tumor virus
MSV	murine sarcoma virus
neo	aminoglycoside 3′-phosphotransferase II gene (encodes resistance to neomycin)
nos	nopaline synthase gene
orf	open reading frame

cos	cohesive sites (λ phage genome)	*ori*	origin of replication (for replication in *E. coli* unless indicated otherwise)
dHfr	dihydrofolate reductase gene	P	promoter
EBNA-1	Epstein–Barr virus nuclear antigen	p10	p10 protein
EBV	Epstein–Barr virus	pA	polyadenylate
EK	enterokinase	PBP	basic protein promoter
EMC	encephalomyocarditis virus	PH	polyhedrin
Ery	erythromycin resistance	P*lac*	lactose promoter
ETL	early to late	P*mt*	mouse metallothionein I gene promoter
F/R	forward and reverse primers	P*spa*	protein A promoter
GAL1	yeast galactokinase gene	P*tac*	hybrid *trp–lac* promoter
gpt	guanine hypoxanthine phosphoribosyl transferase gene	P*trp*	tryptophan promoter
gst	glutathione S-transferase gene	RBS	ribosome binding site
GUS	β-glucuronidase	Rec Seq	recombination sequences
HAT	hypoxanthine/aminopterin/thymidine	rrnBT1T2	transcription terminators
His	histidine	RSV	Rous sarcoma virus
HSV	herpes simplex virus	Splice	splice sites
Hum	human	Stop	stop codons for translation
hyg	hygromycin resistance	SV40	simian virus 40
IPTG	isopropylthiogalactopyranoside	T1, T2	see rrnBT1T2

Abbreviations

TAA	translation stop signal	*ts*	temperature-sensitive mutation
Tc	tetracycline resistance	VPI	SV40 VPI processing signal
tel	telomeric sequence	Xa	factor Xa cutting site
term	transcriptional terminator	YAC	yeast artificial chromosome
TK	thymidine kinase		
Tn	transposon		

PREFACE

Compiling a compact data book on vectors is not an easy task! The number of vectors described in the literature is vast; however, a book of this size cannot possibly be encyclopedic. Therefore, we have decided at the outset to restrict the information to those vectors which are commercially available. This represents between 230 and 250 different vectors, which have been consolidated into approximately 120 composite maps. Clearly, by limiting the total number of vectors, it is inevitable that some areas of research will be less well served and we are conscious of this fact. However, it has been our aim to provide a reasonably balanced coverage of the subject area commensurate with the size of the book. Although most of the information that has been incorporated into the book is already in the public domain, it has never been collected together in an accessible and user-friendly format.

Peter Gacesa and Dipak P. Ramji

Acknowledgments

We are grateful to our colleagues, especially David Willey, for their constructive comments during the writing of this book and to our families for their forbearance and support. We also acknowledge permission to reproduce information, including vector maps from Amersham International plc, Appligene Inc., New England Biolabs, Boehringer Mannheim, Clontech Laboratories Inc., Invitrogen Corporation Inc., Life Technologies Ltd, NBL Gene Sciences Ltd, New Brunswick Scientific Ltd, Novagen Inc., Pharmingen, Pharmacia Biotech Ltd, Promega Ltd, Sigma Chemical Company Ltd, Stratagene Ltd and USB Corporation Ltd.

Chapter 1 **USER'S GUIDE**

You will obtain maximum benefit from the book if you read this brief section first.

1 Layout of the book

In essence, Chapter 2 *Tables 1–4* list all of the vectors described in the book grouped according to their major functional properties, that is cloning and transcription vectors (*Table 1*), lambda-based vectors and cosmids (*Table 2*), expression vectors (*Table 3*) and specialized vectors (*Table 4*). These tables are cross-referenced to the maps in Chapter 3, the name of the commercial supplier of the vector, a reference to GenEMBL accession numbers (if available) and appropriate further information (usually in scientific journals).

Table 5 in Chapter 2 is an alphanumeric list of vectors indicating which of the more common restriction enzyme sites are available and usable. The table is also cross-referenced to the vector maps in Chapter 3 and to *Tables 1–4* of Chapter 2. In addition, *Table 6* of Chapter 2 correlates some of the less common restriction endonuclease sites with vector names.

The bulk of the book (Chapter 3) comprises composite vector maps divided into two groups, viz. *Maps 1–100* (plasmids, phagemids, cosmids) and *Maps 101–117* (lambda phage-based vectors), together with brief descriptive notes. Within each group, the major vectors are arranged alphanumerically. Derivative vectors are often incorporated into composite maps with a closely related major vector.

There is information in Chapter 4 on both the functional use (*Table 1*) and genotypes (*Table 2*) of *Escherichia coli* host strains, and genotypes of lambda phage vectors (*Table 3*) together with a key to genetic markers (*Table 4*). The names and addresses of the major commercial suppliers of vectors are included (Chapter 5), as is a comprehensive list of references.

2 Ways to find information

Some examples of typical questions are given below, together with a possible route to finding appropriate answers.

1. *How do I find a transcription vector with a T7 promoter and an* EcoRI *cloning site?*
If you require a transcription vector, then go initially to Chapter 2, *Table 1*. Promoters are listed in the fourth column. Select any vector(s) with a T7 promoter, e.g. pGEM-1. The map number is indicated in the penultimate column, e.g. *Map 42*.

this particular example it is *Table 3*. Whereas the map will provide brief notes about the appropriate vector, the tables (*Table 1, 2, 3* or *4*) will provide details such as sequencing primers, whether it is possible to obtain single-stranded DNA, as well as cross-references to the scientific literature and details of commercial suppliers.

3. *Which host strains of* E. coli *overexpress the* LacI *repressor?*
Chapter 4 *Table 1* lists host strains which are most suited for specific functions or have particular characteristics. A suitable *Lac*I overproducing strain might be JM109. *E. coli* host cell genotypes are listed in Chapter 4 *Table 2*.

3 Vector maps

Vector maps have been kept as simple as possible while still conveying essential information. Therefore, in most cases, only the more common isoschizomers of restriction endonucleases are included, or two isoschizomers are used

2. *How do I find a vector with* Bam*HI,* Eco*RI,* Eco*RV and* Xba*I cloning sites?*

Go to Chapter 2 *Table 5* and locate vectors that have this combination of restriction endonuclease sites, e.g. pBC. The two columns on the right will direct you to the relevant map (*17*) in Chapter 3 and to one of *Tables 1–4* of Chapter 2; in interchangeably, e.g. *Sac*I and *Sst*I. Similarly, *Sma*I is used in preference to *Xma*I; both enzymes recognize the same site, although their cutting patterns are different. Restriction sites that occur more than once in the polylinker or elsewhere in the vector are included in parentheses.

Chapter 2 **CATALOG OF VECTORS**

Table 1. General cloning and *in vitro* transcription vectors

Vector	*E. coli* host	Marker	Promoters	ssDNA	Sequence primers	Size (kb)	GenBank	Supplier[a]	Map	Ref.
ColE1	C600	ColE1				6.6		Sig	2	1, 2
M13mp8 (9, 18/19, BM20/21)	JM101	BW		M13	M13(F/R)	7.2		BRL, NBS, Pha, USB	3	3–6
pACYC184	RR1	Cm, Tc				4.2		NBL	12	7, 8
pAT153	HB101	Ap, Tc				3.7	L08853	NBL, NBS	20	9
pBR322	HB101	Ap, Tc			Pst(cw/ccw) Sal(cw/ccw)	4.4	L08654	NBL, NBS, Pha, Pro, USB	20	10, 11
pBR325	HB101	Ap, Cm, Tc				6	L08855	BRL, NBL, NBS	20	12
pBR328	HB101	Ap, Cm, Tc				4.9	L08858	Boe, NBL	20	13
pCF20	DH5α	Ap, Tc	T7	f1	T7	4.9		USB	29	14
pEMBL (18, 19+/−)	XL-1Blue	Ap, BW		f1		4	L08863-6	Boe	37	15, 16
pGEM-1/2	JM109	Ap	SP6, T7		SP6, T7	2.9		Pro	42	17
pGEM-3/4	JM109	Ap	SP6, T7		SP6, T7	2.9		Pro	42	17
pGEM-3Z (4Z)	JM109	Ap, BW	SP6, T7		SP6, T7	2.7	X65304/5	Pro	42	17
pGEM-3Zf(+/−)	JM109	Ap, BW	SP6, T7	f1	SP6, T7	3.2	X65306/7	Pro	43	17
pGEM-5Zf(+/−)	JM109	Ap, BW	SP6, T7	f1	SP6, T7	3	X65308/9	Pro	43	17
pGEM-7Zf(+/−)	JM109	Ap, BW	SP6, T7	f1	SP6, T7	3	X65310/1	Pro	43	17
pGEM-9Zf(−)	JM109	Ap, BW	SP6, T7	f1	SP6, T7	2.9	X65312	Pro	43	17

pGEM-11Zf(+/−)	JM109	Ap, BW	SP6, T7	f1	SP6, T7	3.2	X65313/4	Pro	43	17
pGEM-13Zf(+/−)	JM109	Ap, BW	SP6, T7	f1	SP6, T7	3.2	X65315	Pro	43	17
pMB9	RR1	Tc				5.3		Sig	2	18
pMOS Blue	MOSBlue	Ap, BW	T7	f1	T7	2.9		Ame	62	19, 20
pPhagescript SK	XL-1Blue	BW	T3, T7	M13	M13(F/R), SK, KS, T3, T7	7.3	L08874	Str	65	21
pSHlox-1	E. coli	Ap	SP6, T7	f1	SP6, T7	3.8		Nov	114	22
pSL1180 (1190)	NM522	Ap, BW		M13	M13(F/R)	3.4		Pha	73	23
pSP6/T3	E. coli	Ap	SP6, T3		SP6, T3	2.9		BRL	74	24
pSP6/T7-19	E. coli	Ap	SP6, T7		SP6, T7	2.9		BRL	74	24
pSP64	JM109	Ap	SP6		SP6	3	X65327	Boe, NBS, Pro	75	25
pSP64 polyA	JM109	Ap	SP6		SP6	3	X65328	Pro	75	26
pSP65	JM109	Ap	SP6		SP6	3	X65329	Boe, NBS, Pro	75	25
pSP70	JM109	Ap	SP6, T7		SP6, T7	2.4	X65330	Pro	76	27
pSP71	JM109	Ap	SP6, T7		SP6, T7	2.4	X65331	Pro	76	27
pSP72	JM109	Ap	SP6, T7		SP6, T7	2.5	X65332	Pro	76	27
pSP73	JM109	Ap	SP6, T7		SP6, T7	2.5	X65333	Pro	76	27
pSPT18 (19, BM20/21)	JM109	Ap	SP6, T7		SP6, T7	3.1		Boe	78	3
pT3T7*lac*	JM109	Ap, BW	T3, T7		T3, T7	2.7		Boe	83	3
pT7-0	JM109	Ap	T7		T7	4.1		USB	85	14
pT7-1(2)	JM109	Ap	T7		T7	2.8		USB	85	14
pT7T3D	JM109	Ap, BW	T3, T7	f1	M13(F/R), T3, T7	2.9		Pha	89	28
pT7Blue	Novablue	Ap, BW	T7	f1	T7	2.9		Nov	86	22

Continued

Table 1. General cloning and *in vitro* transcription vectors, *continued*

Vector	E. coli host	Marker	Promoters	ssDNA	Sequence primers	Size (kb)	GenBank	Supplier[a]	Map	Ref.
pT7T3 18U (19U)	NM522	Ap, BW	T3, T7	f1	M13(F/R), T3, T7	2.9	L08953	Pha	89	28
pTRXN +/−	XL-1Blue	Ap	T7	f1	T7, pTRXN (+/−)	2.9		USB	88	14
pTZ18R, 19R, 18U, 19U	NM522	Ap, BW	T7	f1	M13(F/R), T7	2.9		NBS, USB, Pha	89	29, 30
pUR222	DH1	Ap, BW				2.7	L09145	Boe, NBS	92	31

[a]See Chapter 5 for details of suppliers.

Abbreviations: Ap, ampicillin resistance; BW, blue/white screening; Cm, chloramphenicol resistance; ColE1, colicin E1 immunity; F/R, forward and reverse primers; Pst(cw/ccw), *Pst*I clockwise/counterclockwise primers; Sal(cw/ccw), *Sal*I clockwise/counterclockwise primers; Tc, tetracycline resistance.

Table 2. Lambda phage/cosmid vectors

Vector	Type	Host	Marker	Promoters	ssDNA	Sequence primers	Expression	Size (kb)	GenBank	Supplier[a]	Map	Ref.
λ2001	R	*E. coli*	*spi*/P2					43		NBL	101	32
λBlueMid	MI	*E. coli*	Ap, BW	T3, T7	M13	T3, T7		43		Clo	102	33
λDASHII	MR	*E. coli*	*spi*/P2	T3, T7		T3, T7		41.9		Str	105	34
λDR2	MI	*E. coli*, human	Ap, Hyg	RSV-LTR, TK		DR2	E	45.7	U02428	Clo	103	35, 36

λEMBL3	R	E. coli	spi/P2					42.2	U02425/53	NBS, Pro, Str	104	37, 38
λEMBL4	R	E. coli	spi/P2					42.2		NBS, Pro, Str	104	37, 38
λEMBL12	R	E. coli	spi/P2					43		Boe	104	3
λEXlox(+)	MI	E. coli	Ap	SP6, T7	f1	SP6, T7, T7 gene10	P, F			Nov	114	22
λFIXII	MR	E. coli	spi/P2	T3, T7		T3, T7		41.9		Str	105	34
λGEM-2	MI	E. coli		SP6, T7		SP6, T7		43.8		Pro	106	17
λGEM-4	MI	E. coli	Ap	SP6, T7		SP6, T7		46.2		Pro	106	17
λGEM-11/12	MR	E. coli		SP6, T7		SP6, T7		43		Pro	107	17, 38, 39
λgt10	I	E. coli	cI repressor			gt10(F/R)		43.3		Ame, BRL, Boe, Inv, NBS, Clo, Pro	108	40, 41
λgt11	I	E. coli	BW			gt11(F/R)	P, F	43.7		Ame, BRL, Boe, Inv, NBS, Clo, Pha, Pro	109	42
λgt11D	I	E. coli	BW			gt11(F/R)	P, F	42.8		Pha	109	28
λgt11 Sfi-Not	I	E. coli	BW			gt11(F/R)	P, F	43.7		Pro	109	17
λgt18/19	I	E. coli	BW			gt11(F/R)	P, F	43.2			110	43
λgt20/21	I	E. coli	BW			gt11(F/R)	P, F	42.7			111	43
λgt22/23	I	E. coli	BW			gt11(F/R)	P, F	42.7			111	43
λgt22A	MI	E. coli	BW	SP6, T7		gt11(F/R), SP6, T7	P, F	42.8		BRL	111	24
λgtWES.λb	I	E. coli						40.3		BRL	112	24
λMax1	MI	E. coli, yeast	Ap, ura3	T7, GAL1, f1 10		T7	E	43.9		Clo	113	44

Continued

Table 2. Lambda phage/cosmid vectors, *continued*

Vector	Type	Host	Marker	Pro-moters	ssDNA	Sequence primers	Ex-pression	Size (kb)	GenBank	Supplier[a]	Map	Ref.
λMOSElox	MI	*E. coli*	Ap	SP6, T7	f1	SP6, T7, T7 gene10	P, F	42.7	M54945	Ame	114	45
λMOSSlox	MI	*E. coli*	Ap	SP6, T7	f1	SP6, T7		42.8		Ame	114	45
λPOP6	MI	*E. coli*, mam.	Ap, Hyg, *cl* repressor	RSV-LTR, TK			E			Inv	115	46
λSHlox-1	MI	*E. coli*	Ap	SP6, T7	f1	SP6, T7			M37056	Nov	114	22
λYES	MI	*E. coli*, yeast	Ap, *ura3*	*lac*, GAL1			E	47.8		Clo	116	44
λZAP II	MI	*E. coli*	Ap, BW	T3, T7, *lac*	f1	M13(F/R), SK, KS, T3, T7	P, F	40.8		Str	117	47, 48
pHC79	C	*E. coli*	Ap, Tc					6.5		Boe, BRL, NBS	50	49
pWE15	C	*E. coli*	Ap, *neo*	T3, T7		T3, T7		8.2		Clo	93	50–52
pWE16	C	*E. coli*	Ap, *dHfr*	T3, T7		T3, T7		8.2		Clo	93	50–52
SuperCos	C	*E. coli*, mam.	Ap, *neo*	T3, T7, SV40		T3, T7		7.6		Str	98	50–52

[a]See Chapter 5 for details of suppliers.

Abbreviations, see *Table 1* plus: C, cosmid; *dHfr*, dihydrofolate reductase gene; E. eukaryotic expression; F, fusion protein; GAL1, yeast galactokinase gene; Hyg, hygromycin B resistance (*hph* gene); I, insertion vector; *lac, lac* promoter; LTR, long terminal repeat; mam., mammalian cells; MI, multifunctional insertion vector; MR, multifunctional replacement vector; *neo, neo* gene; P, prokaryotic expression; R, replacement vector; RSV, Rous sarcoma virus; TK, thymidine kinase gene; *ura3, ura3* gene.

Table 3. Expression vectors

Vector	Host	Marker	Pro-moters	ssDNA	Sequence primers	Ex-pression	Size (kb)	GenBank	Supplier[a]	Map	Ref.
pAX4a b c +/−, pAX5 +/−	E. coli	Ap	lac	f1	pAX1/2	P, F	6.2		USB	15	53–55
pBC SK(+/−) [KS(+/−)]	E. coli	Cm, BW	T3, T7	f1	T3, T7	P, F	3.4		Str	17	21
pBluescript II KS+/− (SK+/−)	E. coli	Ap, BW	T3, T7	f1	M13(F/R), SK, KS, T3, T7	P, F	3		Str	21	47, 56
pBS (+/−)	E. coli	Ap, BW	T3, T7	f1	M13(F/R), T3, T7	P, F	3.2	L08782/3	Str	21	47, 56
pBTac1 (2)	E. coli	Ap, Tc	tac			P	4.6		Boe	22	57
pBTrp2	E. coli	Ap, Tc	trp			P	5.2		Boe	23	58, 59
pCDM8	E. coli, mam.		T7, CMV	M13	T7, pCDM8R	E	4.4		Clo, Inv	26	44, 46
pcDNAI	E. coli, mam.		SP6, T7, CMV	M13	SP6, T7	E	4		Inv	26	46
pcDNAII	E. coli	Ap, BW	SP6, T7, lac	f1	M13(F/R), SP6, T7	P, F	3		Inv	27	46
pcDNA3	E. coli, mam.	Ap, neo	SP6, T7, CMV	f1	SP6, T7	E	5.4		Inv	67	46
pcDNAI/Amp	E. coli, mam.	Ap	SP6, T7	M13	SP6, T7	E	4.8		Inv	26	46
pCDV1	E. coli, mam.	Ap	SV40			E	3.1		Pha	28	28, 60

Continued

Table 3. Expression vectors, *continued*

Vector	Host	Marker	Pro-moters	ssDNA	Sequence primers	Ex-pression	Size (kb)	GenBank	Supplier[a]	Map	Ref.
pCITE	*E. coli*	Ap	T3, T7	f1	M13(F/R), T3, T7, CITE	P, F	3.8		Nov	31	61, 62
pDR2	*E. coli*, human	Ap, Hyg	RSV		EBV(R)	E	10.7	U02428	Clo	35	35
pDR540	*E. coli*	Ap	*tac*			P	4.1		Pha	36	63
pEBVHis A,B,C	*E. coli*, mam.	Ap, Hyg	TK, RSV		EBV(R)	E	10.3		Inv	68	64–67
pET (see map*)	*E. coli*	Ap, Km*	T7*, *lac**	f1*		P, F*	5*		Nov	38	22, 68
pEUK-C1	*E. coli*, mam.	Ap	SV40			E	4.9	U02429	Clo	39	69, 70
pEX1, 2, 3	*E. coli*	Ap	λP_R			P, F	5.8		Boe	40	71, 72
pEXlox(+)	*E. coli*	Ap	SP6, T7	f1		P, F	4		Nov	114	22
pEZZ18	*E. coli*	Ap	*spa, lac*	f1	M13(F)	P, F	4.6	M74186	Pha	41	73
pGEMEX-1 (-2)	*E. coli*	Ap	SP6, T3	f1	SP6, T3, T7 gene10	P, F	4	X65317/8	Pro	44	17
pGEX vectors	*E. coli*	Ap	*tac*			P, F	4.9		Pha	47	74, 75
pKK223-3	*E. coli*	Ap	*tac*			P	4.6	M77749	Pha	53	76
pKK233-2	*E. coli*	Ap	*trc*			P	4.6	U02438	Clo	55	77, 78
pKK388-1	*E. coli*	Ap, Tc	*trc*			P	5.1	U02444	Clo	56	77, 78
pMAL-p2	*E. coli*	Ap, BW	*tac*	M13	*malE*	P, F	6.7		Bio	57	79–81
pMAM	*E. coli*, mam.	Ap, *gpt*	MMTV-LTR			E	7.6	U02443	Clo	58	82, 83

Name	Host	Markers	Promoter	f1	Primers	E/P	Size	GenBank	Source	Ref	Refs
pMAMneo	E. coli, mam.	Ap, neo	MMTV-LTR, RSV-LTR			E	8.4	U02432	Clo	58	82, 83
pMC1871	E. coli	Tc			M13(F)	P, F	7.5		Pha	59	84, 85
pMEX5/6/7/8	E. coli	Ap	tac	f1	pMEX(F/R)	P	3.6		USB	61	54, 55
pMSG	E. coli, mam.	Ap, gpt	MMTV			E	7.6		Pha	63	82
pNH16a	E. coli	Ap	lac			P			NBL, Str	64	21, 86, 87
pNH8a, 18a, 46a	E. coli	Ap	lac, tac			P			NBL, Str	64	21, 86, 87
pPL-lambda	E. coli	Ap	λP_L			P	5.2		Pha	66	28
pRc/CMV	E. coli, mam.	Ap, neo	SP6, T7, CMV	f1	SP6, T7	E	5.5		Inv	67	46
pRc/RSV	E. coli, mam.	Ap, neo	RSV	f1		E	5.2		Inv	67	46
pREP10	E. coli, mam.	Ap, Hyg	TK, RSV		pREP(F), EBV(R)	E	9.5		Inv	68	64–67
pREP4	E. coli, mam.	Ap, Hyg	TK, RSV		pREP(F), EBV(R)	E	10.2		Inv	68	64–67
pREP7	E. coli, mam.	Ap, Hyg	TK, RSV		pREP(F), EBV(R)	E	9.5		Inv	68	64–67
pREP8	E. coli, mam.	Ap, His	RSV, SV40		pREP(F), EBV(R)	E	11.8		Inv	68	64–67
pREP9	E. coli, mam.	Ap, neo	TK, RSV		pREP(F), EBV(R)	E	10.5		Inv	68	64–67
pRIT2T	E. coli	Ap	λP_R			P, F	4.3		Pha	69	88
pRSET A,B,C	E. coli	Ap	T7	f1	T7, pRSET(R)	P, F	2.9	X54202-9	Inv	70	46
pSE280	E. coli	Ap	trc			P	3.9		Inv	71	46

Continued

Table 3. Expression vectors, *continued*

Vector	Host	Marker	Pro-moters	ssDNA	Sequence primers	Ex-pression	Size (kb)	GenBank	Supplier[a]	Map	Ref.
pSE380	*E. coli*	Ap	*trc*			P	4.5		Inv	71	46
pSE420	*E. coli*	Ap	*trc*			P	4.6		Inv	71	46
pSG5	*E. coli*, mam.	Ap	T7, SV40	M13	T7	E	4.1		Str	72	89
pSL301	*E. coli*	Ap	T3, T7, *lac*	f1	M13(F/R), T3, T7	P	3.2		Inv	27	23, 46
pSPORT1	*E. coli*	Ap, BW	SP6, T7, *lac*	f1	M13(F/R), SP6, T7	P, F	4.1		BRL	77	24
pSVK3	*E. coli* mam.	Ap	SV40, T7	f1	T7	E	3.9		Pha	81	28
pSVL	*E. coli* mam.	Ap	SV40			E	4.9		Pha	82	70, 90
pTOPE	*E. coli*	Ap	T7	f1	T7, T7 gene10	P, F	4		Nov	86	22
pTrc99A	*E. coli*	Ap	*trc*			P	4.2		Pha	87	91
pTrcHis	*E. coli*	Ap	*trc*		pTrcHis(F)	P, F	4.4		Inv	71	46
pUC vectors	*E. coli*	Ap, BW			M13(F/R)	P, F	2.7	L08959	NBL, NBS, USB	91	3, 92–94
pXPRS +/−	*E. coli*, mam.	Ap	SV40	f1	pXPRS(+/−)	E	3.8		USB	94	95
pXT1	*E. coli*, mam.	Ap, Tc, *neo*	TK, MMLV-LTR			E	10.4		Str	95	96, 97

[a]See Chapter 5 for details of suppliers.

Abbreviations, see *Tables 1* and *2* plus: CITE, CAP-independent translation enhancer; CMV, cytomegalovirus; EBV, Epstein–Barr virus; *gpt, gpt* gene; His, histidinol; Km, kanamycin resistance; λP_R, λP_R promoter; *malE, malE* gene; MMLV, Moloney murine leukemia virus; MMTV, murine mammary tumor virus; λP_L, λP_L promoter; (R), reverse primer; *spa,* Protein A gene; *tac, tac* promoter; *trc, trc* promoter; *trp,* tryptophan promoter.

Table 4. Specialized vectors

Vector	Host	Marker	Promoters	ssDNA	Sequence primers	Expression	Size (kb)	GenBank	Supplier[a]	Map	Ref.
BacPAK6	BV	β-Gal, BW					130		Clo	1	98
p2Bac	E. coli, BV	Ap	PH, P10		PH(F/R), P10	E	7.1		Inv	4	99, 100
pAc360	E. coli, BV	Ap	PH		PH(F/R)	E, F	9.8		Inv	4	99, 100
pAcAB3	E. coli, BV	Ap	PH, P10(x2)		PH(F/R), P10	E	10		Par	6	101
pAcGP67 A,B,C	E. coli, BV	Ap	PH		PH(F/R)	E, F	9.8		Par	5	101
pAcJP1	E. coli, BV	Ap	39 kDa protein			E	11		Par	6	101
pAcMP1	E. coli, BV	Ap	Basic protein			E	10		Par	6	101
pAcMP2/3	E. coli, BV	Ap	Basic protein			E	10		Par	6	101
pAcUW1	E. coli, BV	Ap	P10		P10	E	4.5		Par	7	101
pAcUW21	E. coli, BV	Ap	PH, P10	f1	PH(F/R), P10	E	9.3		Par	8	101
pAcUW31	E. coli, BV	Ap	PH, P10	M13	Bac1/2, PH(F/R), P10	E	8.5	U02452	Clo	9	98, 102
pAcUW41	E. coli, BV	Ap	PH, P10	f1	PH(F/R), P10	E	7.1		Par	10	101
pAcUW42/43	E. coli, BV	Ap	PH, P10	f1	PH(F/R), P10	E	7.1		Par	10	101
pAcUW51	E. coli, BV	Ap	PH, P10	f1	PH(F/R), P10	E	5.9		Par	11	101

Continued

Catalog of Vectors

Table 4. Specialized vectors, *continued*

Vector	Host	Marker	Pro-moters	ssDNA	Sequence primers	Ex-pression	Size (kb)	GenBank	Supplier[a]	Map	Ref.
pADβ	*E. coli*, mam.	Ap, β-Gal	Adenovirus			β-Gal	7.1	U02442	Clo	13	103–105
pALTER-1	*E. coli*	Ap(s), Tc, BW	SP6, T7	f1	SP6,T7, M13(F/R)		5.7	X65334	Pro	14	17
pBacPAK8/9	*E. coli*, BV	Ap	PH	M13	PH(F/R), Bac1/2	E	5.5	U02450	Clo	16	98
pBI101 (.2 and .3)	Plants	Km, GUS			GUS		12		Clo	18	106–108
pBI121	Plants	Km, GUS	CaMV 35S		GUS	GUS	13		Clo	18	106–108
pBI221	*E.coli*, plants	Ap, GUS	CaMV 35S			GUS	5.7		Clo	18	106–108
pBIN19	Plants	Km, β-Gal					10			18	106–108
pBlueBacHis A,B,C	*E. coli*, BV	Ap, β-Gal, BW	PetI, PH		PH(F/R)	E,F	10		Inv	4	99, 100
pBlueBacIII	*E. coli*, BV	Ap, β-Gal, BW	PetI, PH		PH(F/R)	E	10		Inv	4	99, 100
pBPV	*E. coli*, mam.	Ap	Pmt			E	13		Pha	19	109, 110
pCaMVCN	*E. coli*, plants	Ap, *cat*	CaMV 35S				4.2		Pha	24	111, 112
pCAT-Basic	*E. coli*, mam.	Ap, *cat*					4.4	X65322	Pro	25	17
pCAT-Control	*E. coli*, mam.	Ap, *cat*	SV40				4.8	X65321	Pro	25	17

pCAT-Enhancer	E. coli, mam.	Ap, *cat*					4.6	X65319	Pro	25	17
pCAT-Promoter	E. coli, mam.	Ap, *cat*	SV40				4.5	X65320	Pro	25	17
pCH110	E. coli, mam.	Ap, β-Gal	GPT, SV40				7.1	U02445	Pha	30	113, 114
pCM7	E. coli, mam.	Ap, *cat*					4.1		Pha	32	103, 115
pCMVEBNA	E. coli, mam.	Ap	CMV			EBNA-1	5.5		Clo, Inv	33	35
pCMVβ	E. coli, mam.	Ap	CMV			β-Gal	7.2		Clo	13	103–105
pCRII	E. coli	Ap, Km, BW	SP6, T7	f1	M13(F/R), SP6, T7		3.9		Inv	34	116–118
pG + host 4	Gm +ve	Ery			M13(F/R), SK, KS, T3, T7		3.8		Apl	48	119
pG + host 5	E. coli, Gm +ve	Ery			M13(F/R), SK, KS, T3, T7		5.2		Apl	48	119
pG + host 6	E. coli, Gm +ve	Ap, Ery			M13(F/R), SK, KS, T3, T7		6.7		Apl	48	119
pGEM-*luc*	E. coli	Ap, *luc*	SP6, T7	f1	SP6, T7, M13(F/R)		4.9	X65316	Pro	45	17
pGEM-T	E. coli	Ap, BW	SP6, T7	f1	SP6, T7, M13(F/R)		3		Pro	46	17
pGL2-Basic	E. coli, mam.	Ap, *luc*		f1	GL(1/2)		5.6	X65323	Pro	49	17
pGL2-Control	E. coli, mam.	Ap, *luc*	SV40	f1	GL(1/2)		6	X65324	Pro	49	17

Continued

Table 4. Specialized vectors, *continued*

Vector	Host	Marker	Pro-moters	ssDNA	Sequence primers	Ex-pression	Size (kb)	GenBank	Supplier[a]	Map	Ref.
pGL2-Enhancer	E. coli, mam.	Ap, *luc*		f1	GL(1/2)		5.9	X65325	Pro	49	17
pGL2-Promoter	E. coli, mam.	Ap, *luc*	SV40	f1	GL(1/2)		5.8	X65326	Pro	49	17
pGUSN358-S	E. coli, plants	Ap, GUS		M13	GUS	GUS	5	U02441	Clo	18	106–108
pHph0	E. coli, mam.	Ap, Hyg					3.8		Boe	51	120–122
pHph−1 (+1)	E. coli, mam.	Ap, Hyg					3.8		Boe	51	120–122
pHSV106	E. coli, mam.	Ap	TK				7.8		BRL	52	123
pKK232-8	E. coli	Ap, *cat*					5.1		Pha	54	124
pMAMneo-CAT	E. coli, mam.	Ap, *neo, cat*	RSV				9.2	U02431	Clo	58	82, 83
pMAMneo-LUC	E. coli, mam.	Ap, *neo, luc*	RSV				10	U02448	Clo	58	82, 83
pMC1neo	E. coli, mam.	Ap, *neo*	TK			E	3.8		Str	60	125
pMC1neo polyA	E. coli, mam.	Ap, *neo*	TK			E	3.8		Str	60	125

pMSG-CAT	E. coli, mam.	Ap, cat, gpt	MMTV		E	8.4		Pha	63	82
pNASSβ	E. coli	Ap, β-Gal				6.5	U02433	Clo	13	45, 103, 104
pNEO	E. coli	Ap, neo				5.5		Pha	32	126, 127
pSV-β-galactosidase	E. coli, mam.	Ap, β-Gal	SV40			6.8	X65335	Pro	79	17
pSV2neo	E. coli, mam.	Ap, neo				5.7	U02434	Clo	80	35
pSVβ	E. coli, mam.	Ap, β-Gal	SV40			6.9	U02435	Clo	13	45, 103, 104
pT3/T7luc	E. coli	Ap, luc	T3, T7, lac	T3, T7		4.7	U02437	Clo	84	128
pTKβ	E. coli, mam.	Ap, β-Gal	TK		β-Gal	7.5		Clo	13	45, 103, 104
pUB110	B. subtilis	Km				4.5		Sig	90	129
pUC4K	E. coli	Ap, Km, neo		4				Pha	32	93
pVL941	E. coli, BV	Ap	PH	PH(F/R)	E	9.8		Par	4	101
pVL1392/3	E. coli, BV	Ap	PH	PH(F/R)	E	9.6		Inv	4	99,100
pYAC2 (3/4/5)	E. coli, yeast	Ap, ura3				12	U01086	Sig	96	130, 131
pYACneo	E. coli, mam., yeast	Ap, neo, ura3				16		Clo	96	130, 131

Continued

Table 4. Specialized vectors, *continued*

Vector	Host	Marker	Pro-moters	ssDNA	Sequence primers	Ex-pression	Size (kb)	GenBank	Supplier[a]	Map	Ref.
pYES2	*E. coli*, yeast	Ap, *ura3*	T7, GAL1	f1	T7	E	5.9		Inv	97	46, 132
YEp24	*E. coli*, yeast	Ap, Tc, *ura3*					7.7		Bio	99	133–135
YIp5	*E. coli*, yeast	Ap, Tc, *ura3*					5.5		Bio	100	133–135

[a]See Chapter 5 for details of suppliers.

Abbreviations, see *Tables 1–3* plus; Ap(s), see *Map 14*; BV, baculovirus; CaMV, cauliflower mosaic virus; *cat*, chloramphenicol acetyltransferase gene; Ery, erythromycin resistance; β-Gal, β-galactosidase; Gm +ve, Gram-positive bacteria; GPT, *gpt* promoter; GUS, β-glucuronidase; *luc*, luciferase gene; PH, polyhedrin; Pmt, promoter from mouse metallothionein I gene.

Table 5. Common restriction sites in vectors

Vector	ApaI	AvaI	BamHI	BglII	BstXI	ClaI	EcoRI	EcoRV	HindIII	KpnI	NcoI	NheI	NotI	NruI	PstI	SacI	SalI	SmaI	SpeI	SphI	XbaI	XhoI	XmaIII	Map	Table
ColE1		•			•	•	•							•				•						2	1
λ2001			•				•		•							•					•	•		101	2
λBlueMid							•		•															102	2
λDASHII			•				•		•							•					•	•		105	2
λDR2			•																		•			103	2
λEMBL3/4			•				•										•							104	2
λEMBL12			•				•									•	•	•			•			104	2
λEXlox(+)	•						•		•							•					•			114	2
λFIXII							•										•				•	•		105	2
λGEM-2/4							•									•					•	•		106	2
λGEM-11			•				•									•					•	•		107	2
λGEM-12			•				•						•			•					•	•		107	2
λgt10							•																	108	2
λgt11							•																	109	2
λgt11D							•						•											109	2
λgt11*Sfi-Not*							•						•											109	2
λgt18/19							•										•							110	2
λgt20/21							•										•				•			111	2
λgt22/23							•						•			•	•				•			111	2

Continued

Table 5. Common restriction sites in vectors, *continued*

Vector	ApaI	AvaI	BamHI	BglII	BstXI	ClaI	EcoRI	EcoRV	HindIII	KpnI	NcoI	NheI	NotI	NruI	PstI	SacI	SalI	SmaI	SpeI	SphI	XbaI	XhoI	XmaIII	Map	Table
λgt22A							•						•				•	•			•			111	2
λgtWES.λb							•										•							112	2
λMax1							•														•	•		113	2
λMOSElox	•						•		•								•							114	2
λMOSSlox	•						•		•								•							114	2
λPOP6			•																		•			115	2
λSHlox-1	•						•		•								•							114	2
λYES							•															•		116	2
λZAPII			•				•		•	•	•		•			•	•	•	•	•	•	•		117	2
M13mp8/9			•				•		•						•		•	•						3	1
M13mp12/13			•				•		•						•	•	•	•			•			3	1
M13mp18/19			•				•		•	•					•	•	•	•		•	•			3	1
M13 BM20/21	•		•				•	•	•	•	•	•	•		•	•	•	•	•	•	•		•	3	1
p2Bac	•		•	•			•		•		•					•	•		•		•			4	4
pAc360			•				•		•								•	•			•			4	4
pAcAB3			•	•												•					•			6	4
pAcGP67 A,B,C			•	•			•				•		•								•	•		5	4
pAcJP1			•																		•			6	4
pAcMP1			•																		•			6	4
pAcMP2/3			•	•			•						•								•	•		6	4

Vector		
pAcUW1	7	4
pAcUW21	8	4
pAcUW31	9	4
pAcUW41	10	4
pAcUW42/43	10	4
pAcUW51	11	4
pACYC184	12	1
pALTER-1	14	4
pAT153	20	1
pAX4a b c +/−	15	3
pAX5 +/−	15	3
pBacPAK8/9	16	4
pBC SK +/− (KS(+/−))	17	3
pBI101 (.2 and .3)	18	4
pBIN19	18	4
pBlueBacHis	4	4
pBlueBacIII	4	4
pBluescript II KS +/− (SK +/−)	21	3
pBPV	19	4
pBR322	20	1
pBR325	20	1
pBR328	20	1
pBS (+/−)	21	3
pBTac1 (2)	22	3
pBTrp2	23	3

Continued

Catalog of Vectors

Table 5. Common restriction sites in vectors, *continued*

Vector	ApaI	AvaI	BamHI	BglII	BstXI	ClaI	EcoRI	EcoRV	HindIII	KpnI	NcoI	NheI	NotI	NruI	PstI	SacI	SalI	SmaI	SpeI	SphI	XbaI	XhoI	XmaIII	Map	Table
pCAT-Basic									•						•		•			•	•			25	4
pCAT-Control															•	•					•			25	4
pCAT-Enhancer							•								•	•					•			25	4
pCAT-Promoter			•												•		•			•	•			25	4
pCDM8					••				•				•		•						••	••		26	3
pcDNA3	•		•		••		•	•	•	•			•								•	•		67	3
pcDNAI			•		••		•	•	•				•		•						•	•		26	3
pcDNAI/Amp			•		••		•	•	•				•								•	•		26	3
pcDNAII	•		•		••		•	•	•				•		•	•			•	•	•	•	••	27	3
pCDV1									•															28	3
pCF20						•	•																	29	1
pCH110									•	•														30	4
pCITE		•	•	•		•		•	•			•			•	•	•			•	•	•	•	31	3
pCRII	•	•	•		••		••	•	•				•			•		•			•	•		34	4
pDR2			•													•					•			35	3
pDR540			•																					36	3
pEBVHis A,B,C			•			•			•	•												•		68	3
pEMBL (18, 19+/−)			•				•		•	•					•	•	•	•			•			37	1
pET vectors (see map)		•	•			•	•	•	•	•	•		•			•	•				•	•	•	38	3
pEUK-C1			•													•	•		•		•	•		39	3

Vector		
pEX1, 2, 3	40	3
pEXlox(+)	114	3
pEZZ18	41	3
pG⁺host 4	48	4
pG⁺host 5	48	4
pG⁺host 6	48	4
pGEM-1/2	42	1
pGEM-3/4	42	1
pGEM-3Z (4Z)	42	1
pGEM-3Zf(+/−)	43	1
pGEM-5Zf(+/−)	43	1
pGEM-7Zf(+/−)	43	1
pGEM-9Zf(−)	43	1
pGEM-11Zf(+/−)	43	1
pGEM-13Zf(+/−)	43	1
pGEM-luc	45	4
pGEM-T	46	4
pGEMEX-1 (-2)	44	3
pGEX-1λT	47	3
pGEX-2T (-2TK -3X)	47	3
pGEX-4T (-5X)	47	3
pGL2-Basic	49	4
pGL2-Control	49	4
pGL2-Enhancer	49	4
pGL2-Promoter	49	4
pHC79	50	2

Continued

Table 5. Common restriction sites in vectors, *continued*

Vector	ApaI	AvaI	BamHI	BglII	BstXI	ClaI	EcoRI	EcoRV	HindIII	KpnI	NcoI	NheI	NotI	NruI	PstI	SacI	SalI	SmaI	SpeI	SphI	XbaI	XhoI	XmaIII	Map	Table
pHph0									•								•							51	4
pHph+1																		•						51	4
pHph−1		•																•		•				51	4
pHSV106									•	•							•	•		•				52	4
pKK223-3							•		•						•		•	•						53	3
pKK232-8			•						•						•		•	•						54	4
pKK233-2							•		•		•				•		•	•						55	3
pKK388-1							•		•		•	•			•	•		•			•			56	3
pMAL-p2			•				•		•						•		•	•			•			57	3
pMAM												•					•	•				•		58	3
pMAMneo												•					•					•		58	3
pMAMneo-CAT												•										•		58	4
pMAMneo-LUC												•												58	4
pMB9			•	•			•		•								•							2	1
pMC1871			••												••		••	•						59	3
pMC1neo																	•					•		60	4
pMC1neo polyA																	•					•		60	4
pMEX5/6		•				•	•	•	•						•		•			•				61	3
pMEX7/8	•	•	•			•	•	•	•	•	•				•		•	•		•				61	3
pMOS Blue		•	•				•	•	•	•					•	•	•	•	•	•	•			62	1

pMSG		63	3
pMSG-CAT		63	4
pNH8a		64	3
pNH16a		64	3
pNH18a		64	3
pNH46a		64	3
pPhagescript SK		65	1
pRc/CMV		67	3
pRc/RSV		67	3
pREP4		68	3
pREP7		68	3
pREP8		68	3
pREP9		68	3
pREP10		68	3
pRIT2T		69	3
pRSET		70	3
pSE280		71	3
pSE380		71	3
pSE420		71	3
pSG5		72	3
pSHlox-1		114	1
pSL301		27	3
pSL1180 (1190)		73	1
pSP6/T3		74	1
pSP6/T7-19		74	1
pSP64		75	1

Continued

Table 5. Common restriction sites in vectors, *continued*

Vector	ApaI	AvaI	BamHI	BglII	BstXI	ClaI	EcoRI	EcoRV	HindIII	KpnI	NcoI	NheI	NotI	NruI	PstI	SacI	SalI	SmaI	SpeI	SphI	XbaI	XhoI	XmaIII	Map	Table
pSP64 polyA		•	•				•		•						•	•	•	•			•			75	1
pSP65		•	•				•		•						•	•	•	•			•			75	1
pSP70/71				•		•	•	•	•													•		76	1
pSP72/73			•		•	•	•	•	•	•					•	•	•	•		•	•	•		76	1
pSPORT1			•				•		•	•			•		•	•	•	•	•	•	•	•		77	3
pSPT18/19		•	•				•		•	•					•	•	•	•		•	•			78	1
pSPT BM20/21	•	•	•				•	•	•	•	•				•	•	•	•			•	•		78	1
pSVK3	•						•		•						•	••	•	•			•	•		81	3
pSVL							•		•						•	•		•			•	•		82	3
pT3T7*lac*		•	•				•		•						•	•	•	•		•	•			83	1
pT7-1/2			•				•		•						•	•	•	•			•			85	1
pT7T3D			•				•		•				•		•	•		•		•	•	•		89	1
pT7Blue		•	•				•	•	•	•					•	•	•	•	•	•	•			86	1
pT7T3 18U (19U)			•				•		•	•					•	•	•	•		•	•			89	1
pTOPE		•	•		••		•	•	•						•	•	•		•		•	•		86	3
pTrc99A			•				•		•	•	•				•	•	•	•			•			87	3
pTrcHis		•	•	•			•		•	•					•							•		71	3
pTRXN +/−			•		•		•		•	•							•	•		•				88	1
pTZ18R/18U/19R/19U			•				•		•	•					•	•	•	•			•	•		89	1
pUB110				•																				90	4

pUC8/9	91	3
pUC12/13	91	3
pUC18/19	91	3
pUC-BM20/21	91	3
pUR222	92	1
pVL941	4	4
pVL1392/3	4	4
pWE15 (16)	93	2
pXPRS +/−	94	3
pXT1	95	3
pYAC2 (3/4/5)	96	4
pYACneo	96	4
pYES2	97	4
SuperCos	98	2
YEp24	99	4
YIp5	100	4

Table 6. Special restriction sites in vectors

Restriction enzyme	Vector
*Aat*II	pSPORT1, pGEM-5Zf/7Zf, pGEM-T, pSL301, pSE280/380/420
*Afl*II	pSL1180/1190, p2Bac, pSL301, pSE280/380/420
*Apo*I	pUC18/19, pTZ18R/19R/18U/19U, pNEB193
*Asc*I	pNEB193, p2Bac
*Ava*I	pHph + 1, ColE1, M13 BM20/21, pBR325, pCITE, pCRII, pEMBL, pET, pGEM 3/4, 3Z/4Z, 3Zf; pHpH-1, pHSV-106, pMEX5/6/7/8, pMOSBlue, pSE280, pSL1180/1190, pSP64/65/64 polyA, pSPT18/19/BM20/BM21, pT3/T7lac, pT7Blue, pTOPE, pTrcHis A,B,C, pUC 18/19/BM20/BM21, pNEB193
*Ava*III	pMEX7/8, pSL301, pSE380/420
*Avr*II	λGEM11, pSL1180/1190, pSE280
*Bal*I	pSL301, pSE380/420, pCITE
*Ban*I	pT7Blue
*Ban*II	pT3T7lac, pT7-0, pUC18/19, pTZ18R/19R/18U/19U, pNEB193, pCF20
*Bbr*PI	pSL1180/1190
*Bbv*II	pSE280
*Bcl*I	pNEO, pSL1180/1190, pSL301, pSE280
*Bfr*I	pUCBM20/21, M13BM20/21, pSPTBM20/21, pSL1180/1190
*Bsm*I	pSL301, pSE280/380/420
*Bsi*WI	p2Bac
BspMI	pMOSBlue, pUC18/19, pTZ18U/19U/18R/19R, pNEB193, pSE420, pSL301
*Bsp*MII	pSPORT1, pSL1180/1190, pSL301, pSE380/420
*Bss*HII	pNEB193, pSL1180/1190, p2Bac, pSL301, pSE280
*Bst*1107I	p2Bac
*Bst*BI	pBacPAK8/9, pSL1180/1190, pSL301, pSE280/380/420, pTrcHis A,B,C; pRSET A,B,C, pGEM 7Z(f)

*Bst*EII	pMEX8/9, pSL301, pSE280
*Bsu*36I	BacPAK6, pSL1180/1190, pSL301, pSE280/380/420
*Dra*II	pG + host4/5/6, pBluescript, pBC, Phagescript, pSE280, pSL301, pSE380/420
*Dra*III	pSE380/420
*Eco*47III	pSL1180/1190, pSL301, pSE380/420
*Hpa*I	pPL-Lambda, pSL1180/1190, pSL301, pSE280/380/420
*Mlu*I	pSPORT1, pUCBM20/21, M13BM20/21, pSPTBM20/21, pGL2 vectors, pSL1180/1190, pSL301, pSE280
*Nae*I	pSE280/380/420
*Nar*I	pSL301, pSE280
*Nde*I	pT7Blue, pCITE, pMOSBlue, pGEM-5Zf, pGEM-T, pSL1180/1190, pSL301
*Nhe*I	pSL301, pAT153, pBR322, pBR325, pBR328, pGEM 3/4, 3Zf; pGL2 vectors, pMAM/neo/neo-CAT, YEp24, YIp5
*Nru*I	pSL1180/1190, pSL301, pSE280/380/420, ColE1, pACYC184, pAT153, pAX 4a,b,c/5, pBR322, pBR325, pNH16/18/46a, YIp5
*Nru*II	pSE380/420
*Nsi*I	pMEX7/8, pGEM-5Zf/7Zf/9Zf/11Zf, pGEMEX-1/2, pGEM-*luc*, pGEM-T, pSL1180/1190, pcDNAI/II/I-amp, pCRII, pSL301, pSE280
*Pac*I	pBacPAK8/9, pNEB193
*Pfl*MI	pSL301, pSE280
*Pvu*I	pSV2 neo, pBR325, pBR322
*Pvu*II	pT7-0, pSP70/71/72/73, pREP4/7/10, p2Bac, pRSET A,B,C
*Rsr*II	pSPORT1
*Sac*II	pG + host4/5/6, pBluescript, pBC, Phagescript, pTOPE, pGEM-5Zf, pGEM-T, pBPV, pSL1180/1190, p2Bac, pSL301, pSE280/380/420, pUCBM20/21, M13BM20/21
*Sfi*I	pGEM-11Zf/13Zf, pGEMEX 1/2, pGEM-*luc*, λgt11 Sfi-Not/D, pREP4/7/9/10, pEBV His A,B,C, pSL1180/1190, p2Bac, pSE280/380/420
*Sna*BI	pSPORT1, pYAC3, pSL1180/1190, pSL301, pSE280/380/420
Continued	

Catalog of Vectors

Table 6. Special restriction sites in vectors, *continued*

Restriction enzyme	Vector
*Spl*I	pSPORT1, pSL1180/1190, pSL301, pSE280/380/420
*Srf*I	p2Bac, pSL301
*Sse*8387I	pBacPAK8/9, pT7Blue, pMOSBlue, pUC18/19, pTZ18U/19U/18R/19R, pNEB193
*Ssp*I	pBR322
*Stu*I	pBacPAK8/9, pGEM-luc, pSL1180/1190, p2Bac, pSL301, pSE280/380/420
*Sty*I	pUCBM20/21, M13BM20/21, pSPTBM20/21, pMEX7/8, pBR322, p2Bac
T-cloning site	pCRII, pCITE, pGEM-T, pMOSBlue, pT7Blue(R), pTOPE-1b
*Tth*111I	pSL301
*Xca*I	pSL301
*Xcm*I	pCITE
*Xmn*I	pMAL-p2/c2

Chapter 3 **VECTOR MAPS**

Map 1. BacPAK6 and related vectors.

Baculovirus expression systems are designed for high-level expression of recombinant proteins in insect cells. The most efficient transfection of insect cells is achieved by using replication-deficient forms of the virus, *Autographa californica*.

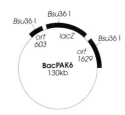

Foreign DNA is inserted into the cloning site of a transfer vector, i.e. pBacPAK8 or pBacPAK9. Recombinant DNA and *Bsu*36I-digested BacPAK6 (a modified form of the viral genome) are co-transfected into host cells. Homologous recombination between the viral genome and the transfer vector results in the restoration of viral replication and the incorporation of foreign DNA into the viral genome.

BacPAK6 contains the *lacZ* gene so that 5-bromo-4-chloro-3-indolyl-β-D-galactoside (X-Gal) may be used to distinguish parental (blue) from recombinant (white) virus plaques.

Map 2. ColE1 and related vectors.

E. coli containing ColE1 vectors produce colicin E1, a protein that inhibits the growth of other bacteria. ColE1 vectors also impart host cell immunity to low levels of colicin E1.

There are unique restriction sites in the *colE1* gene in ColE1.

A related vector, ColE1-Amp (10.9 kb) has unique sites for *Eco*RI and *Sma*I in the colicin-producing gene, and *Bam*HI in the ampicillin gene. Strains carrying the unmodified ColE1-Amp vector produce colicin, have colicin immunity and are ampicillin resistant.

Vector pMB9 confers tetracycline resistance and colicin immunity.

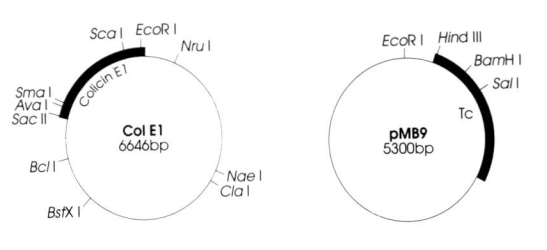

Map 3. M13mp vectors.

The M13-based vectors provide either double-stranded or single-stranded DNA. Ideal for the rapid purification of single-stranded DNA template for dideoxy sequencing.

The *lacZ* gene allows blue/white screening of recombinants (see *Map 1*). A series of M13 vectors with multiple cloning sites (MCS) of different complexity and orientation have been constructed. MCS sequences are identical to the corresponding pUC vectors.

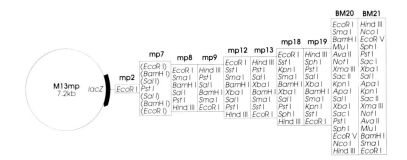

Vector Maps

Map 4. pAc360 and related vectors.

Baculovirus transfer vectors providing both transcriptional and translational initiation sequences for expression of fusion proteins (see also *Map 1*, BacPAK6).

pAc360 is used to produce recombinant proteins with a polyhedrin N-terminus by insertion of DNA at the *Bam*HI site, 36 bp downstream of the polyhedrin (PH) translational start codon.

pBlueBacHis A, B and C express proteins in each of three reading frames with an N-terminal peptide histidine tag for one-step purification by metal-chelate chromatography. The His tag is removed using enterokinase. pBlueBacHis has an MCS suitable for each of three reading frames. The *Autographa californica* nuclear polyhedrosis virus (AcMNPV) early to late (ETL) promoter and *lacZ* gene allow blue/white screening of recombinants.

p2Bac permits expression of two recombinant proteins from the same construct from the AcMNPV PH and p10 enhancer/promoter sequences.

pBlueBacIII has the *lacZ* gene for blue/white screening of recombinants.

pVL941 is similar to pAc360.

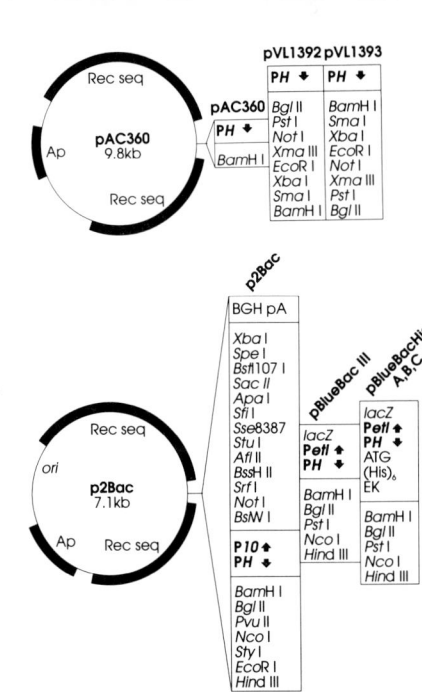

Map 5. pAcGP67A.

Baculovirus vector designed for the expression of recombinant proteins as a gp67 signal peptide fusion product.

The signal ensures secretion of the recombinant protein and facilitates purification, particularly if protein-free insect culture medium is used.

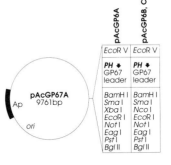

Map 6. pAcJP1 and related vectors.

Baculovirus vectors designed for protein expression.

pAcMP1 and pAcMP2/3 are designed for expression of foreign genes in the late phase of virus infection, i.e. prior to the very late phase when the PH and the P10 genes are expressed.

pAcJP1 allows expression during the delayed early and late phases of the infection cycle. Expression is reduced compared to PH- and p10-promoter-driven vectors, but post-translational modifications are more readily accomplished in this period.

pAcAB3 allows the simultaneous expression of three foreign genes during the very late phase of infection.

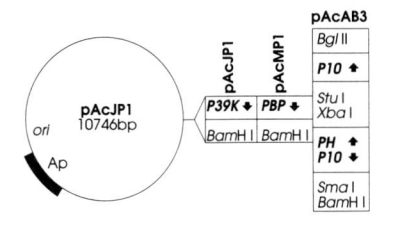

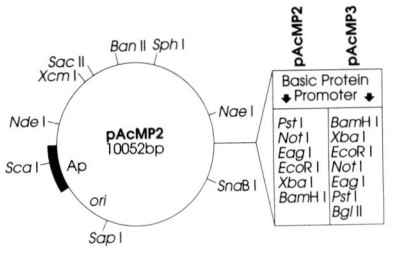

Map 7. pAcUW1.

Baculovirus expression vector.

Expression of genes inserted into the *Bgl*II site is driven by a p10 gene promoter.

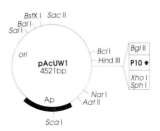

Map 8. pAcUW21.

A baculovirus expression vector with a p10 gene promoter and SV40 transcription termination sequences inserted upstream of the complete polyhedrin gene.

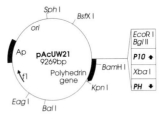

Map 9. pAcUW31.

A baculovirus transfer vector with two strong AcMNPV promoters for the simultaneous high-level expression of two genes.

The vector PH and p10 promoters have similar strengths and are active in the same phase of infection.

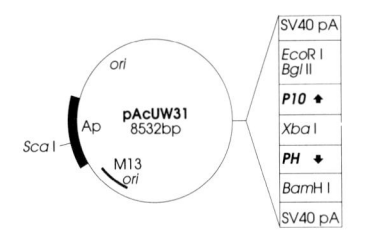

Flanking AcMNPV sequences allow transfer of the expression cassette to the viral polyhedrin (see *Map 1*, BacPAK6 for details).

These vectors have a pUC origin of replication, an M13 origin for single-stranded DNA production and an ampicillin gene for selection in *E.coli*.

Map 10. pAcUW41 and related vectors.

pACUW41, 42 and 43 have the PH and the p10 gene promoters in tandem.

A copy of the SV40 transcription termination sequences has been inserted between the promoters to prevent readthrough.

The insert DNA should have an ATG codon situated less than 100 bp from the cloning site.

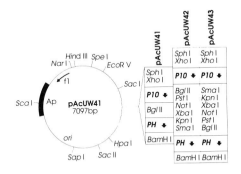

Map 11. pAcUW51.

Designed for the simultaneous expression of two cloned proteins.

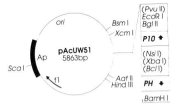

Map 12. pACYC184.

pACYC184 can be used together with, or as an alternative vector to, pBR322 and derivatives.

pBR322 (and its derivatives) and pACYC184 have different but compatible replicons and therefore can coexist in the same cell. This allows two recombinant plasmids to be maintained and expressed simultaneously.

DNA insertion and gene inactivation are possible at unique sites in the chloramphenicol or tetracycline genes.

A related vector, pACYC177, has similar properties to pACYC184 but encodes kanamycin rather than chloramphenicol resistance.

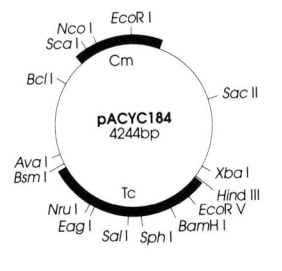

Map 13. pADβ and related vectors.

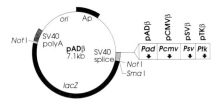

pADβ, pCMVβ, pSVβ and pTKβ are reporter vectors designed for the expression of β-galactosidase in mammalian cells. Transcription is from either adenovirus (pADβ), cytomegalovirus (pCMVβ), herpes simplex virus thymidine kinase (pTKβ) or SV40 early (pSVβ) promoters.

pNASSβ is a promoterless mammalian reporter vector that allows cloning and testing of promoters using β-galactosidase expression.

These vectors can be used for optimizing electroporation conditions, as reference plasmids for reporter gene constructs, as 'enhancer-trap' vectors or for analyzing *cis*-acting elements and *trans*-acting factors.

All vectors contain an SV40 RNA splice site, polyadenylation signal and the full-length *E. coli lacZ* gene. The *lacZ* gene is flanked by *Not*I restriction sites to facilitate excision or replacement with an alternative gene. The *Sma*I site is absent in pTKβ.

Map 14. pALTER-1.

Designed for site-directed mutagenesis experiments.

The vector has a nonfunctional ampicillin gene but an ampicillin-repair oligonucleotide is used to convert the plasmid to ampicillin resistant at the same time as the cloned insert is mutated. Therefore, screening for ampicillin resistance ensures that a high frequency of mutated cloned inserts are obtained.

May also be used for general purpose cloning, *in vitro* transcription (T7 or SP6 promoters), blue/white screening of recombinants and production of single-stranded DNA.

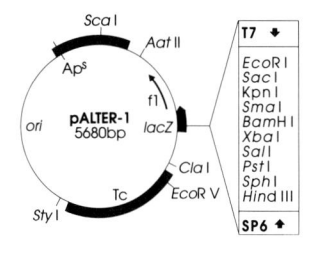

Map 15. pAX vectors.

Expression vectors which enable genes to be cloned in each of three reading frames (a, b, c).

Genes are expressed as fusion proteins with β-galactosidase and collagen fragment N-terminal sequences. Endoproteinase Xa digestion, following purification of the product, releases native protein.

An f1 origin of replication (+ or − orientation) allows rescue of either strand of DNA.

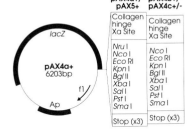

pAX4a+/− pAX5+	pAX4b+/− pAX4c+/−
Collagen hinge Xa Site	Collagen hinge Xa Site
Nru I Nco I Eco RI Kpn I Bgl II Xba I Sal I Pst I Sma I	Nco I Eco RI Kpn I Bgl II Xba I Sal I Pst I Sma I
Stop (x3)	Stop (x3)

Map 16. pBacPAK8 and related vectors.

Transfer vectors for baculovirus-mediated expression of recombinant proteins in insect cells.

Flanking AcMNPV sequences allow transfer of the expression cassette to the viral polyhedrin (see *Map 1*, BacPAK6 for details).

High-level expression of a cloned gene is driven by the strong AcMNPV PH promoter and the *Pac*I site provides translational stop codons in all three reading frames.

These vectors have a pUC origin of replication, an M13 origin for single-stranded DNA production and an ampicillin gene for selection in *E.coli*.

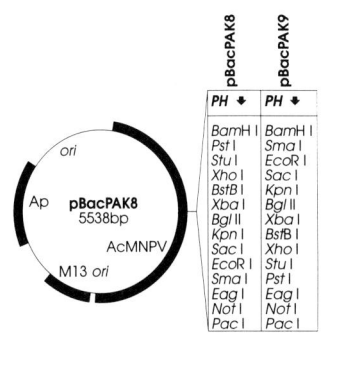

pBacPAK8	pBacPAK9
PH ↓	PH ↓
BamH I	BamH I
Pst I	Sma I
Stu I	EcoR I
Xho I	Sac I
BstB I	Kpn I
Xba I	Bgl II
Bgl II	Xba I
Kpn I	BstB I
Sac I	Xho I
EcoR I	Stu I
Sma I	Pst I
Eag I	Eag I
Not I	Not I
Pac I	Pac I

Map 17. pBC phagemids.

Derived from pBluescript II vectors by replacement of the original ampicillin gene with one for chloramphenicol resistance.

Used for *in vitro* transcription (T7 or T3 promoters), rescue of single-stranded DNA, generation of nested deletions using exonuclease III, direct sequencing and fusion protein expression. The *lacZ* gene allows blue/white screening of recombinants.

The polylinker is in the opposite orientation in the pBC KS (+) phagemid.

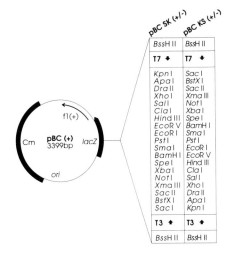

Map 18. pBIN19 and related vectors.

pBIN19 is a low-copy-number vector maintained in host cells with kanamycin.

pBI101 is a reporter vector derived from pBIN19 for testing promoter function in plant host cells. Insertion of a suitable promoter at the polylinker results in β-glucuronidase (GUS) expression. All three reading frames, relative to the polylinker, are available, i.e. pBI101.1, pBI101.2 and pBI101.3.

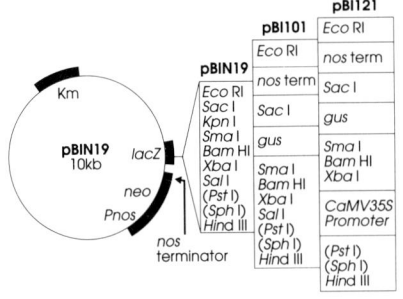

The ATG internal to the polylinker *Sph*I site may cause problems with promoter sequences inserted upstream, i.e. at the *Bam*HI site. The polyadenylation signal from the *Agrobacterium* Ti plasmid nopaline synthase (*nos*) gene has been inserted downstream of the GUS gene.

pBI121 has been derived from pBI101 by insertion of the cauliflower mosaic virus (CaMV) 35S promoter upstream of the GUS gene which ensures high-level expression of GUS in tobacco cell transformants.

pBI221 has been constructed by insertion of the *Hind*III–*Eco*RI fragment of pBI121 (CaMV 35S promoter, GUS gene and nopaline synthase terminator) into the high-copy-number plasmid pUC19 which facilitates large-scale preparation of DNA, e.g. for electroporation of protoplasts.

pGUSN358-S (map not shown) is similar to pBI221 except that there is an M13 insert to facilitate isolation of single-stranded templates and the cryptic GUS *N*-glycosylation site has been removed. This latter modification ensures that GUS is not inactivated in the endoplasmic reticulum by *N*-glycosylation. Therefore, pGUSN358-S is ideal for secretory and vacuolar targeting studies, in addition to studies of the expression of nuclear and cytoplasmic proteins. The initiating ATG of the GUS gene is in a unique *Nco*I site (CCATGG) which lies within a consensus translational initiator.

Map 19. pBPV.

Designed for stable expression of genes in mammalian cells without integration into the host genome. The plasmid contains the entire genome of bovine papilloma virus and is maintained episomally at 20–150 copies per cell.

Expression is controlled by the enhancer from the long terminal repeat of the Moloney murine sarcoma virus and the promoter (Pmt) from the mouse metallothionein I gene. Translation starts at the first ATG in the cloned insert and SV40 splice and polyadenylation signals ensure efficient expression.

Transformed cells have altered morphology and loss of growth control, which is used as a dominant selectable marker. Recommended mammalian cell lines include murine C127 or 3T3 cells or Fisher rat fibroblasts (FR3T3).

This vector has a pBR322 replicon and an ampicillin gene for growth in *recA* strains of *E. coli*.

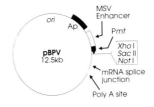

Map 20. pBR322 and related vectors.

General-purpose vectors for subcloning DNA fragments.

pAT153 is a derivative of PBR322 which lacks ancilliary sequences involved in the control of copy number and mobilization. Both pBR325 (not shown) and pBR328 are also derivatives of pBR322 which contain an additional chloramphenicol resistance gene.

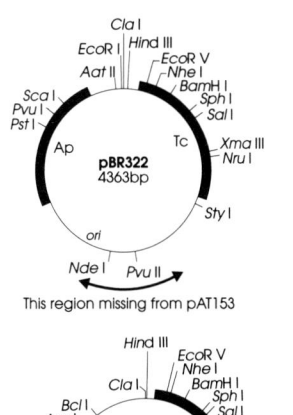

Map 21. pBS(+) and related vectors.

For general cloning, *in vitro* transcription from either strand (T7 or T3 promoters) and rescue of sense or antisense single-stranded DNA by a helper phage. The *lacZ* gene allows blue/white screening of recombinants.

An inducible *lac* promoter upstream from the *lacZ* gene allows the production of fusion proteins. In addition, these vectors can be used directly for dideoxy sequencing of cloned DNA using M13 reverse or universal primers or T7/T3 primers.

pBluescript II SK and KS polylinkers are in opposite orientations.

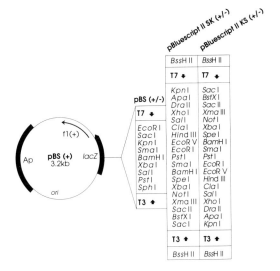

49

Map 22. pB*Tac*.

Used for protein expression in *E. coli* under the control of the strong *trp-lac* (*tac*) promoter.

pB*Tac*2 is similar to pB*Tac*1 but has an ATG initiation codon in the polylinker. Strong ribosomal RNA transcription terminators have been introduced downstream of the multiple cloning sites in both vectors.

The *tac* promoter is regulated by the Lac repressor protein in *lacI^q* hosts (e.g. JM101, JM105, JM107, JM109) but can be induced with IPTG.

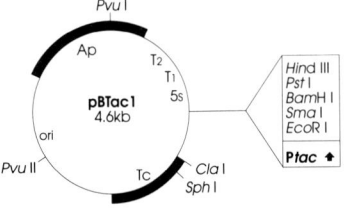

Map 23. pB*Trp*2.

Designed for regulated expression of cloned genes in *E. coli*.

The plasmid contains the tightly regulated tryptophan (*trp*) promoter, a ribosome binding site and an ATG translation initiation codon (as part of the *Nco*I site) within the polylinker.

This vector should be used in a *trp* R⁺ *E. coli* host, e.g. C600. Transcription is repressed by tryptophan but induced by 3-β-indolylacrylic acid.

Map 24. pCaMVCN.

Designed as a positive control for monitoring the efficiency of electroporation into plant cells.

The plasmid encodes chloramphenicol acetyltransferase (*cat*) expressed from the CaMV 35S promoter.

The vector is functional in most mono- and dicotyledonous plants.

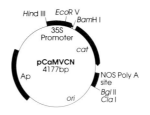

Map 25. pCAT.

These vectors are used to assay promoter or enhancer activity in transfected mammalian cells.

The organization of these vectors is similar to the pGL2 series (*Map 49*) except that they contain the *cat* reporter gene, which encodes chloramphenicol acetyltransferase, instead of *luc*.

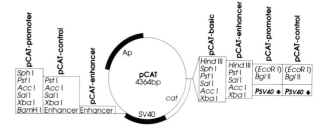

Vector Maps

Map 26. pCDM8 and related vectors.

Insert DNA (up to 7 kb) is expressed transiently in mammalian cells from the strong CMV promoter. SV40 and polyoma origins of replication allow episomal replication in cells expressing the SV40 large T antigen or latently infected with polyoma virus.

Cloning sites are flanked by T7 and/or SP6 promoters.

A ColEI origin of replication ensures maintenance in *E. coli* and the M13 replicon allows production of single-stranded DNA.

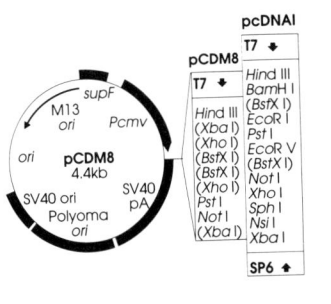

Map 27. pcDNAII and related vectors.

For cDNA cloning, subcloning, production of nested deletions and generation of fusion proteins.

The polylinker in pcDNAII contains 19 unique sites, whereas pSL301 has a superlinker of 46 unique sites. Both vectors allow blue/white selection of recombinants and production of single-stranded DNA.

In vitro transcription of cloned inserts can be produced from flanking promoters in both pcDNAII (SP6/T7) and pSL301 (T3/T7).

pSL301 contains translational initiation sequences and an inducible *lac* promoter for regulated transcription in *E.coli* strains expressing the *lacI^q* repressor.

The pSL301 polylinker is:
NcoI, BstEII, Tth111I, PflMI, BsmI, BstBI, EcoRI, BspMI, AatII, SalI, Bsu36I, SmaI, XmaI, ApaI, Eco0109I, BamHI, BclI, BglII, BssHI, EcoRV, Eco47III, SphI, NheI, XbaI, Asp718, KpnI, HpaI, SnaBI, XcaI, BspMII, StuI, AvrII, NsiI, NdeI, NotI, EagI, PstI, BbeI, NarI, MluI, SplI, NruI, SacII, SacI, PaeR7I, XhoI, SpeI, BalI, SfiI, AflII, HindIII.

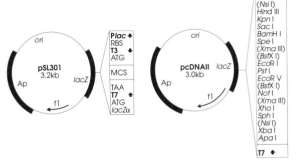

Map 28. pCDV1 (Okayama vector).

For efficient cDNA cloning and subsequent expression in mammalian cells using the method of Okayama and Berg [60].

As the orientation of the cDNA is defined, control signals are correctly positioned to ensure efficient expression in mammalian cells. The pL1 linker has the SV40 origin of replication, early promoter and mRNA splicing sequence positioned upstream of the cDNA. In contrast, the SV40 polyadenylation sequence is located downstream of the cDNA in pCDV1.

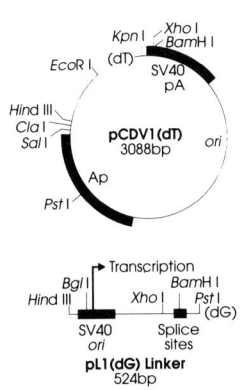

Map 29. pCF20.

General-purpose cloning vector.

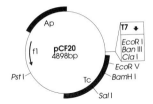

Map 30. pCH110.

Used for estimating transformation/transfection efficiencies and stability by expression of β-galactosidase in *E. coli* (from the *gpt* promoter) or mammalian cells (from the SV40 early promoter).

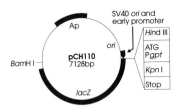

Vector Maps

Map 31. pCITE-2.

Designed for increased translation performance of T7-transcribed RNA of cloned sequences.

The encephalomyocarditis virus (EMC) RNA 5′-non-coding region permits initiation of translation by eukaryotic ribosomes. The CAP-independent translation enhancer (CITE) dramatically increases *in vitro* translation by rabbit reticulocyte lysates and has a similar effect on reporter gene expression in transfected mammalian cells. Cloned sequences fused with the enhancer are transcribed *in vitro* with T7 RNA polymerase without the need for cap analogs.

A C-terminal histidine tag allows affinity purification of translational products.

pCITE is suitable for *Nde*I–*Bam*HI subcloning from pET vectors and *Eco*RI subcloning from λEXlox, λgt11 and λZAP.

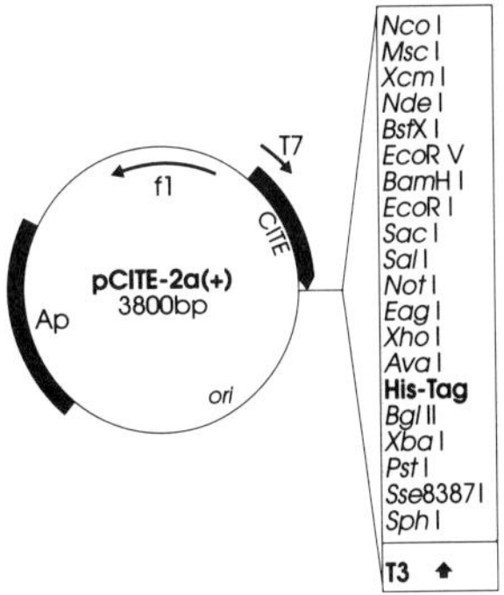

Nco I
Msc I
Xcm I
Nde I
BsfX I
EcoR V
BamH I
EcoR I
Sac I
Sal I
Not I
Eag I
Xho I
Ava I
His-Tag
Bgl II
Xba I
Pst I
Sse8387 I
Sph I

T3

pCITE-2a(+)
3800bp
Ap
f1
T7
CITE
ori

Map 32. pCM7 and related cartridge vectors.

pCM7 has a 780 bp *cat* cartridge which may be excised from the vector using *Hind*III.

pNEO contains the aminoglycoside 3′-phosphotransferase II (*neo*) gene which confers resistance to kanamycin or neomycin in *E. coli* hosts. This gene can be used as a dominant selectable marker in mammalian and plant cells. Also, insertion of the gene into a mammalian transcription unit produces host resistance to G-418.

pUC4K contains the aminoglycoside 3′-phosphotransferase gene derived from transposon *Tn*903 which encodes resistance to kanamycin, neomycin and G-418. The cassette containing the aminoglycoside 3′-phosphotransferase gene can be used to construct new vectors or to disrupt protein coding sequences.

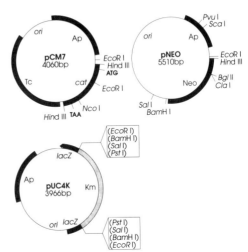

Vector Maps

Map 33. pCMVEBNA.

Control vector for the expression of Epstein–Barr virus nuclear antigen (EBNA-1).

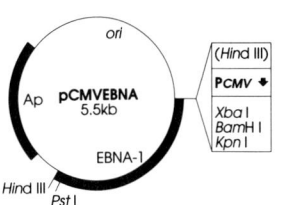

Map 34. pCR™II.

Multifunctional vector for direct cloning of polymerase chain reaction (PCR) products using the T overhang.

SP6 and T7 promoters flank the polylinker for production of sense and antisense RNA transcripts. The *lacZ* gene facilitates blue/white screening of recombinants and expression of PCR products as *lacZ* fusion proteins.

An f1 origin allows production of single-stranded DNA.

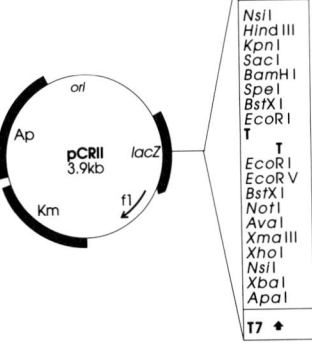

Map 35. pDR2.

An Epstein–Barr virus (EBV)-based vector for stable expression of cDNA in a broad range of human host cells.

pDR2 is maintained episomally at moderate copy numbers (10–20 per cell) and thus provides stable expression when host cells are co-transfected with pCMVEBNA.

Expression of cloned cDNA is driven by the strong Rous sarcoma virus-long terminal repeat (RSV-LTR) promoter.

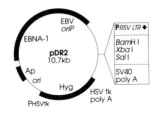

Map 36. pDR540.

Prokaryotic expression vector containing the strong *tac* promoter which is repressed by the *lac* repressor protein in *lacI*^{q} hosts, e.g. *E. coli* JM105. Expression is induced with IPTG.

Transcription can be monitored by assaying galactokinase activity (*galK* gene) using a radiochemical assay.

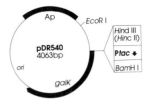

Map 37. pEMBL.

For subcloning and generation of single-stranded DNA (f1 origin). More stable with large cloned inserts than M13 vectors.

The MCS, which is opposite orientations in odd and even numbered vectors, is inserted in the *lacZ* gene to facilitate blue/white screening of recombinants.

Vectors designated + or − have the f1 origin in opposite orientations.

pEMBL18	pEMBL19
EcoR I	Hind III
Sac I	Sph I
Kpn I	Pst I
Sma I	Sal I
BamH I	Xba I
Xba I	BamH I
Sal I	Sma I
Pst I	Kpn I
Sph I	Sac I
Hind III	EcoR I

pEMBL18+ 3960bp — Ap, ori, lacZ, f1

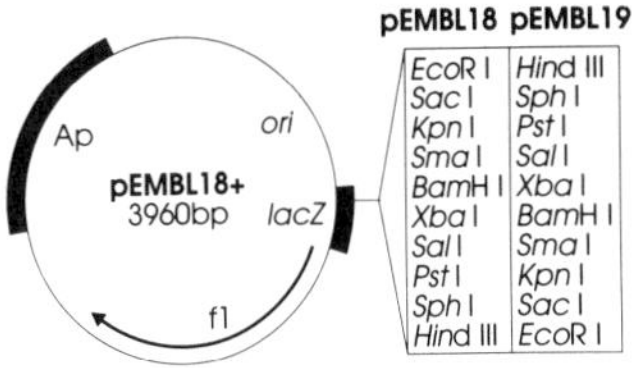

Map 38. pET vectors.

A family of multifunctional prokaryote expression vectors. Features include T7 promoters, f1 origin for the generation of single-stranded DNA and various cleavable N-terminal or C-terminal tags. (See Appendix A: pET vector classification.)

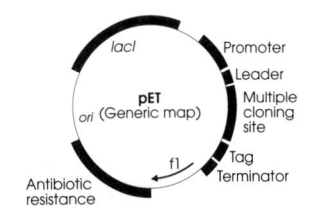

pET (Generic map): lacI, Promoter, Leader, Multiple cloning site, Tag, Terminator, f1, ori, Antibiotic resistance

Map 39. pEUK-C1.

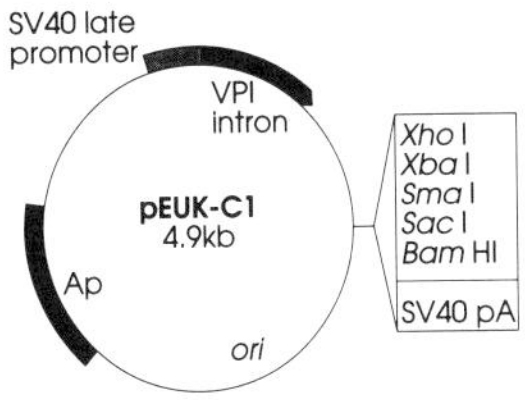

A vector designed for transient expression of cloned genes in mammalian cells.

Transcription is driven from the SV40 late promoter but translation is dependent on an ATG codon in the cloned fragment. Transcripts are spliced and polyadenylated using the SV40-derived processing signals.

The vector has SV40 and pBR322 origins of replication. The absence of pBR322 'poison sequences' results in an increased copy number and high levels of expression in Cos cells.

Map 40. pEX1.

For expression of cDNA libraries as β-galactosidase fusion proteins.

The λcro–E.coli lacZ gene fusion is controlled by the strong λP$_R$ promoter, and the polylinker, situated at the 3' end of lacZ, is present in all three reading frames. Phage fd transcription terminators and synthetic translation 'stop' signals have been inserted downstream of the polylinker.

Host strains must have the temperature-sensitive λcI repressor allele.

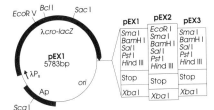

61

Map 41. pEZZ18.

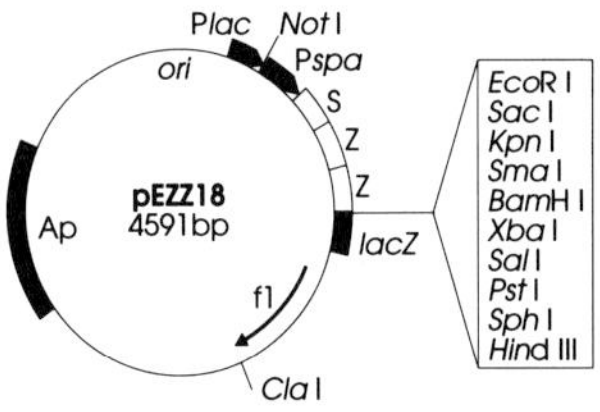

Designed for expression and secretion of Protein A fusion proteins.

The Protein A domain, including a signal sequence, facilitates secretion from *E. coli*, increases the solubility and allows direct purification of the fusion protein by affinity chromatography on IgG Sepharose.

The unique folding properties of the Protein A 14 kDa ZZ peptide results in minimal interference with the folding of recombinant protein into its native conformation.

The *E. coli* host should be an α-complementation *lacZ'* mutant.

Map 42. pGEM-1 and related vectors.

Used as standard cloning vectors and for *in vitro* transcription (T7 or SP6 promoters). pGEM-3Z and pGEM-4Z also contain the *lacZ* gene which allows blue/white screening for recombinants.

pGEM-1 (2865 bp), pGEM-2 (2869 bp), pGEM-3 (2867 bp), pGEM-4 (2871 bp), pGEM-3Z (2743 bp) and pGEM-4Z (2746 bp) are nearly identical except for the orientation and complexity of the multiple cloning site and the absence or presence of a *lacZ* gene.

See also pGEM-3Zf(−) and related vectors (*Map 43*).

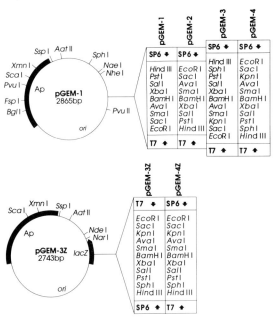

Vector Maps

Map 43. pGEM-3Zf (+ / −) and related vectors.

Used as standard cloning vectors and for *in vitro* transcription (T7 or SP6 promoters). Includes the *lacZ* gene for blue/white screening of recombinants and an f1 origin for production of single-stranded DNA.

pGEM-5Zf (+ / −) and pGEM-7Zf (+ / −) were constructed for the generation of nested deletions using exonuclease III (Erase-a-Base, Promega).

pGEM-9Zf (−) and pGEM-11Zf (+ / −) include restriction sites for direct cDNA cloning strategies or for convenient subcloning of cDNA from the λgt11 *Sfi-Not* vector (Promega).

pGEM-13Zf (+) allows in-frame subcloning and expression of DNA sequences subcloned from the λgt11 *Sfi-Not* vector.

pGEM-3Zf (3199bp), pGEM-5Zf (3003 bp), pGEM-7Zf (3000 bp), pGEM-9Zf (2925 bp), pGEM-11Zf (3223 bp) and pGEM-13Zf (3181 bp) are similar except for the complexity of the multiple cloning site and orientation of the f1 origin.

See also pGEM-1 and related vectors (*Map 42*).

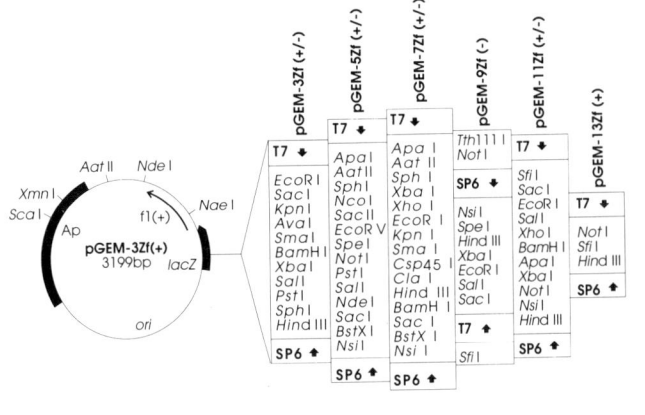

Map 44. pGEMEX-1 and -2.

For *in vitro* transcription (SP6 or T3 promoters) and the production of single-stranded DNA.

DNA is expressed as T7 gene 10 fusion proteins in *E. coli* host strains containing an inducible T7 RNA polymerase, e.g. JM109 (DE3).

pGEMEX-1 (3995 bp) and pGEMEX-2 (3997 bp) differ only in the position of the reading frame.

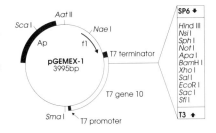

Map 45. pGEM-*luc*.

pGEM-luc is a derivative of pGEM-11Zf(-) which contains the luciferase gene in the centre of the MCS (see *Map 43*).

This vector cassette is designed to be a source of the *luc* gene encoding firefly luciferase.

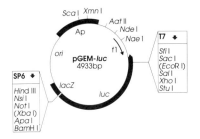

Vector Maps

Map 46. pGEM-T.

The vector is a derivative of pGEM-5Zf(+) (*Map 43*) with single 5'-T overhangs at the insertion site to improve the efficiency of PCR product ligation. This system utilizes the nontemplate-dependent addition of a single deoxyadenosine to the 3' end of PCR products by some thermostable polymerases.

Also useful for general cloning, *in vitro* transcription (T7 or SP6 promoters), blue/white screening of recombinants and production of single-stranded DNA.

The polylinker is specially designed for the generation of nested deletions using exonuclease III.

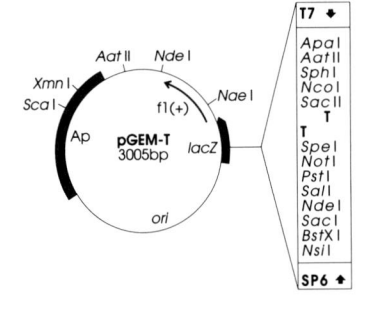

Map 47. pGEX.

For IPTG-inducible, high-level expression (*tac* promoter and *lacI^q* gene) of glutathione S-transferase fusion proteins. Products are purified directly by affinity chromatography on glutathione-Sepharose and the native protein released with thrombin (pGEX-1λT, -2T, -2TK, -4T) or Factor Xa (pGEX-3X, -5X).

Collectively, these vectors allow fusions in all three translational reading frames with respect to the *Eco*RI site. The pGEX-1λT *Eco*RI site is in-frame with the corresponding site in λgt11 for direct expression of inserts from λgt11 expression libraries.

pGEX-2TK has a cAMP-dependent protein kinase recognition site.

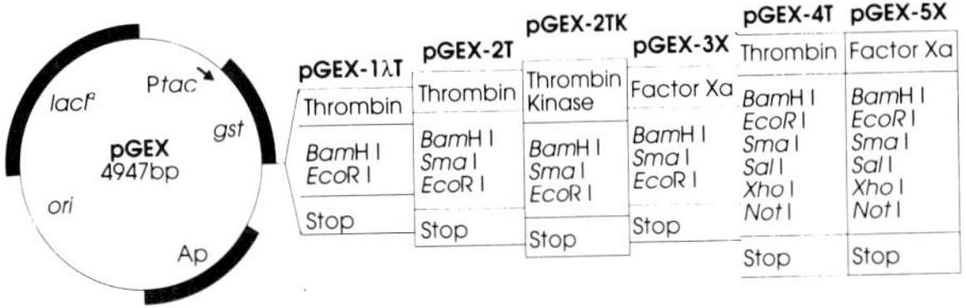

Map 48. pG⁺host.

Broad-host-range vectors for use with *E.coli* and Gram-positive bacteria. Used as delivery vectors for transposon mutagenesis, chromosomal integration and recombinational inactivation of chromosomal genes.

The temperature-sensitive replicon (Ts) is functional at 28°C, but non-functional at 37°C. pG⁺host 5 and 6 have an additional replicon (pBR322) for temperature-independent function in *E. coli*.

Other features include T3 and T7 promoters and markers for erythromycin (pG⁺host 4, 5, 6) and ampicillin (pG⁺host 6).

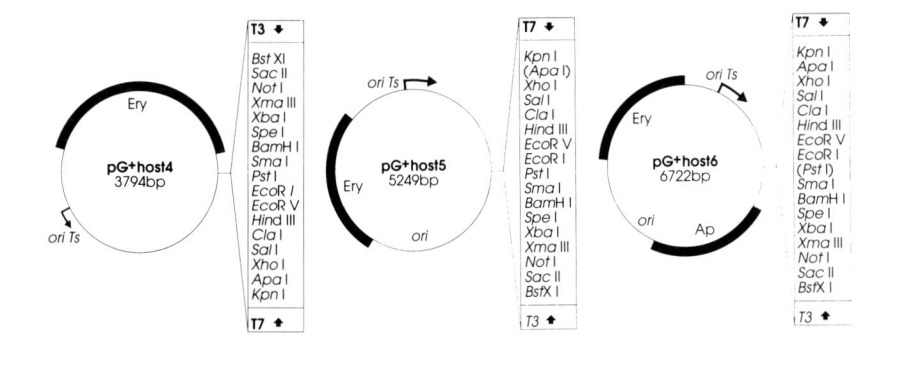

Map 49. pGL2-basic and related vectors.

A family of *luc*-reporter vectors for assaying promoter or enhancer function in transfected mammalian cells.

Vectors lack either a promoter (pGL2-enhancer), enhancer (pGL2-promoter) or both (pGL2-basic). pGL2-control has both an SV40 promoter and enhancer for monitoring transfection efficiency and as a convenient reference for comparison of luciferase activity in different constructs (cf. pCAT vectors; *Map 25*).

Putative promoters should be cloned upstream of the *luc* gene in the correct orientation for effective transcription. Enhancer-containing DNA should be effective when inserted either upstream or downstream of the *luc* gene, and in either orientation.

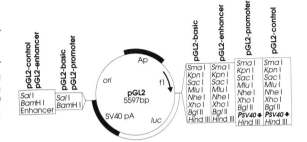

Map 50. pHC79.

A cosmid vector especially suitable for cloning large DNA fragments (up to 40 kb).

The choice of *E. coli* host strain depends on the methylation status of the insert DNA.

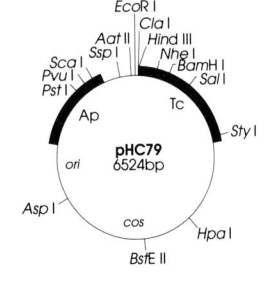

Map 51. pHph.

A family of vectors constructed by insertion of the hygromycin B phosphotransferase (*hph*) gene, which confers resistance to hygromycin, into the *Bam*HI site of different pUC vectors, pUC8 (pHph + 1), pUC9 (pHph0) and pUC12 (pHph − 1).

If cloned DNA contains a suitable promoter, expression of the *lacI–hph* fusion product will confer resistance to hygromycin B.

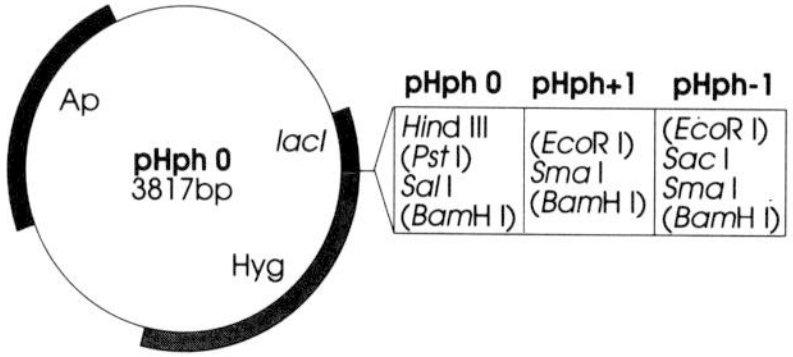

	pHph 0	pHph+1	pHph-1
	Hind III	(*Eco*R I)	(*Eco*R I)
	(*Pst* I)	*Sma* I	*Sac* I
	Sal I	(*Bam*H I)	*Sma* I
	(*Bam*H I)		(*Bam*H I)

Map 52. pHSV106.

For monitoring the activity of cloned promoter or enhancer sequences in transfected cells. Also useful as a marker for DNA transfection of eukaryotic cells.

The vector contains the herpes simplex virus type 1 thymidine kinase (tk) gene inserted at the *Bam*HI site of pBR322.

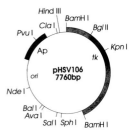

Map 53. pKK223-3.

A prokaryotic expression vector with a strong *tac* promoter, downstream polylinker and strong *rrnB* transcriptional termination signal.

The *tac* promoter is normally repressed in *lacI*^{q} hosts, e.g. *E. coli* JM105, but expression is inducible with IPTG.

Inserts containing a ribosome-binding site and an ATG start codon can be expressed by cloning into the *Pst*I or *Hind*III sites.

Alternatively, the vector ribosome binding site may be used by ligation of the insert at the *Eco*RI or *Sma*I sites. The ATG start codon should be within 10–15 bp of the vector ribosome binding site.

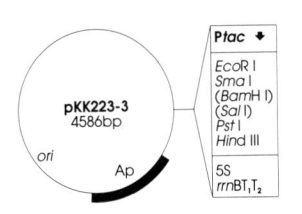

Map 54. pKK232-8.

Designed for cloning and monitoring promoters by assaying expression of the *cat* gene.

Terminators are located upstream of the cloning site and downstream of the *cat* gene to prevent readthrough during transcription. Stop codons in all three reading frames are located between the polylinker and the ATG of the *cat* gene to prevent inadvertent readthrough during translation.

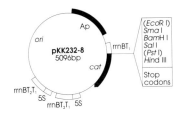

Map 55. pKK233-2.

For high-level expression of cloned genes and cDNAs in *lacI^q* *E. coli* strains (e.g. JM109).

Features include a strong IPTG-inducible *trc* promoter, the *lacZ* ribosome binding site and an ATG initiation codon which is part of the unique *Nco*I cloning site. Transcriptional terminators are situated downstream of the multiple cloning site to prevent readthrough which could disrupt plasmid replication.

73

Map 56. pKK388-1.

Similar to pKK233-2 (*Map 55*) but with an extended cloning site.

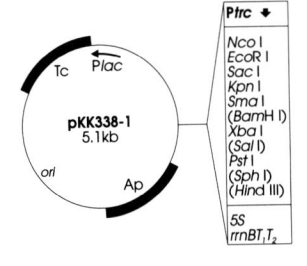

Map 57. pMAL.

For the expression of *malE* (maltose binding protein, MBP) fusion proteins.

Expression of the *malE–lacZα*-fusion protein is controlled by the inducible *tac* promoter. The *lacI^q* gene ensures that the promoter is fully repressed in the absence of IPTG.

Other features include a transcriptional terminator downstream of *lacZα* to prevent readthrough, an M13 origin of replication for the generation of single-stranded DNA, and blue/white screening of recombinants.

pMAL-c is identical to pMAL-p2 except for deletion of the *malE* signal sequence.

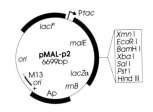

Map 58. pMAM and related vectors.

A family of dexamethasone-inducible mammalian expression vectors, which have a murine mammary tumor virus (MMTV)-LTR promoter.

SV40 splicing and polyadenylation sites provide RNA processing in mammalian cells and the *gpt* gene permits selection of transformants in HAT (hypoxanthine/aminopterin/thymidine) medium. Alternatively, the *neo* gene, driven by the SV40 early promoter, allows selection of transformants in media containing G418 antibiotics.

pMAMneo is designed for high-level expression of eukaryotic genes. The vector contains the RSV-LTR enhancer linked to the promoter which allows controlled, high-level expression of genes in hormonally responsive cell lines (e.g. ATT20, BHK, CHO, GH3, HeLa, mammary gland epithelial, 3T3, U937).

pMAMneo-CAT and pMAMneo-LUC are similar to pMAMneo except that they contain the *cat* and *luc* reporter genes for use as positive controls.

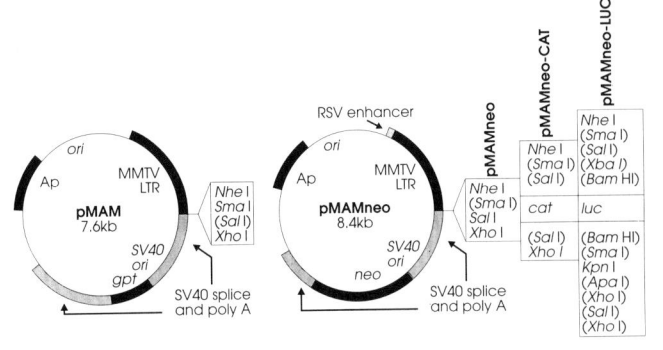

Map 59. pMC1871.

A vector, without promoters, designed for gene expression studies or as a source of the *lacZ* gene cassette.

The *lacZ* gene lacks a ribosome-binding site and codons for the first eight nonessential N-terminal amino acid. Cloning at the unique *Sma*I site generates *lacZ* fusion proteins, provided that inserts contain appropriate promoter and ribosome-binding sequences.

Digestion with *Bam*HI, *Sal*I and *Pst*I allows excision of a *lacZ* gene cartridge. Excision with *Eco*RI results in deletion of the 3' end of *lacZ*; however, the resulting β-galactosidase protein (α-donor) is functional if the C-terminus of the β-galactosidase protein (α-acceptor) is available through intercistronic complementation.

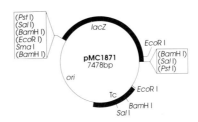

Map 60. pMC1neo and related vectors.

Used for gene targeting and lineage marking in mammalian stem cells.

The *neo* gene, the herpes simplex virus thymidine kinase promoter and the polyoma virus, py F441, enhancer are introduced into an exon of a cloned gene fragment. Homologous recombination confers neomycin resistance on transfected mammalian host cells.

pMC1neo should be used for cloned inserts with a polyadenylation signal. pMC1neo polyA has a vector-encoded polyA signal.

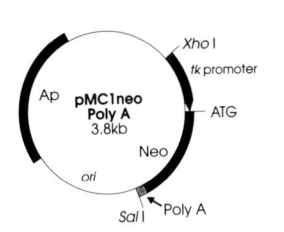

Map 61. pMEX vectors.

Multipurpose vectors designed for cloning, expression, mutagenesis and sequencing (f1 generation of single-stranded DNA).

pMEX6 and pMEX8 are identical to pMEX5 and pMEX7, respectively, except for the orientation of the f1 origin.

Features include transcriptional terminators and stop codons in all three reading frames. Gene expression is controlled by the IPTG-inducible *tac* promoter.

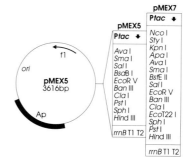

Map 62. pMOS Blue.

Designed for cloning PCR products.

Single 5′-T overhangs at the insertion site to improve the efficiency of PCR product ligation. This system utilizes the nontemplate-dependent addition of a single deoxyadenosine to the 3′ end of PCR products by some thermostable polymerases.

The vector contains a *lacZ* gene for blue/white screening of recombinants, an f1 origin of replication for production of single-stranded DNA and a T7 promoter for *in vitro* RNA synthesis.

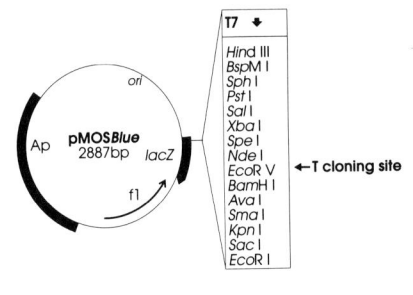

Map 63. pMSG vectors.

For inducible expression of cloned genes in mammalian cells. Genes inserted into the polylinker are expressed from the glucocorticoid-inducible promoter in the MMTV-LTR and are translated from the first ATG codon.

Dexamethasone-responsive cells, (e.g. Chinese hamster ovary (CHO), mouse Ltk⁻ and 3T6, and human HeLa) are required for either transient or stable expression. The *gpt* gene provides a marker for selection of stably transfected cells.

pMSG-CAT is designed as a positive control for pMSG.

Spontaneous deletions may occur as these vectors have sequences homologous to the *E. coli* genome. *E. coli* HB101 is the recommended host strain.

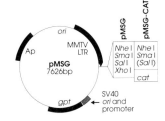

Map 64. pNH8a and related vectors.

Expression of inserts cloned into pNH vectors is regulated by temperature-induced inversion of *lac* or *tac–lac* tandem promoters.

These vectors should be used with the *E.coli* host strains D1210HP and D1210. The phage-derived Int product of the host strain, D1210HP, and the *attB* and *attP* recognition sequences of the plasmid ensure that a heat pulse will cause site-specific conservative recombination of the promoter(s), independently of plasmid replication, and result in expression of cloned DNA.

Both pNH8a and pNH16a have the *galK* gene.

pNH46a uses heat-induced inversion of the polylinker to control expression of cloned genes. Genes that have been inserted in an antisense orientation will invert upon heat pulse treatment and transcription of sense strand will proceed. This minimizes the possibility of expression prior to inversion.

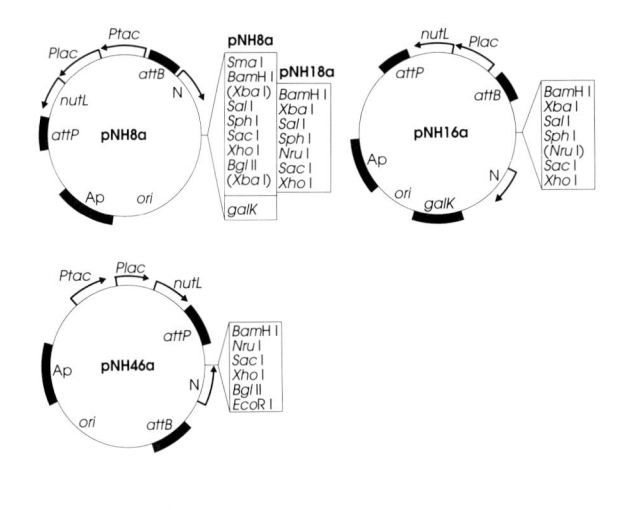

Map 65. pPhagescript.

A modified M13 bacteriophage with blue/white screening of recombinants.

Used as a standard cloning vector, for *in vitro* transcription (T7 or T3 promoters), expression of β-galactosidase fusion proteins, production of single-stranded DNA and creation of nested deletions using exonuclease III.

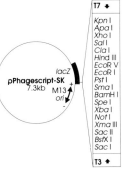

Vector Maps

Map 66. pPL-lambda.

A λP$_L$ promoter containing vector for thermoinducible over-expression of proteins which might be lethal to the host.

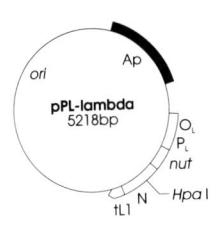

The promoter can be thermoregulated in host strains containing the temperature-sensitive cI857 repressor (e.g. *E. coli* N4830-1). At 29–31°C the cI857 repressor is active, whereas at 42°C it is nonfunctional, i.e. increasing the temperature switches on expression.

In cI$^+$ hosts the promoter can also be induced by nalidixic acid.

Map 67. pRc/CMV and related vectors.

For high-level transient or stable expression of recombinant proteins in mammalian cells.

Constitutive transcription is from the mammalian enhancer/promoter sequences (immediate early gene of human cytomegalovirus (CMV) in pRc/CMV, pcDNA3 and pcDNAI neo, and RSV-LTR in pRc/RSV). Vectors incorporate polyadenylation and transcription termination sequences from bovine growth hormone gene (pRc/CMV and pRc/RSV) and SV40 (pcDNAI neo). Vectors contain an SV40 origin for episomal replication.

The f1 sequence (pRc/CMV, pRc/RSV and pcDNA3) or M13 origin (pcDNAIneo) allows rescue of single-stranded DNA.

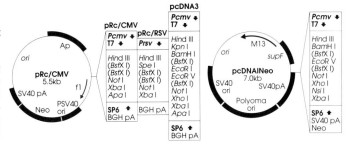

Vector Maps

Map 68. pREP4 and related vectors.

EBV-based vectors for high-level expression of recombinant proteins in a wide range of mammalian cells. These vectors, which are maintained extrachromasomally, have an EBV origin of replication (*oriP*) and the nuclear antigen gene (EBNA-1).

Selection markers are hygromycin (pEBVHis, pREP4, pREP7, pREP10), histidinol (pREP8) or G418 (*neo*)(pREP9). All vectors contain a ColE1 origin of replication and an ampicillin gene for growth and maintenance in *E. coli*.

pREP-CAT is a control vector which expresses chloramphenicol acetyltransferase.

Constitutive expression is from the RSV-LTR enhancer/promoter. Different orientations of the MCS allow high-level expression of sense and antisense transcripts.

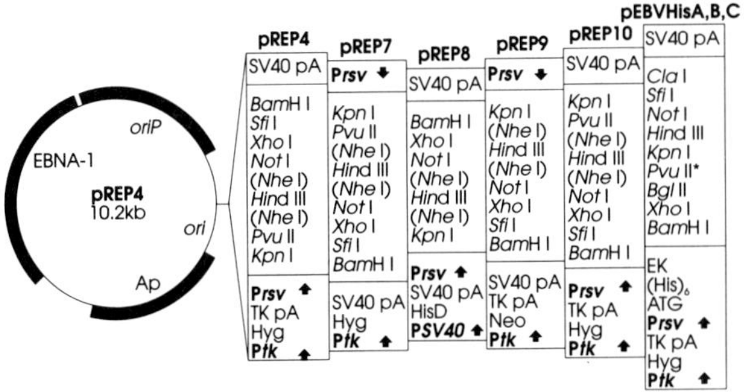

pEBVHis has the MCS in three different reading frames for expression of recombinant protein with an N-terminal histidine tag. This enables one-step purification of product by metal-chelate chromatography before tag removal with enterokinase.

Constitutive expression of EBNA-1 by host cell lines results in a 100-fold increase in transfection. A vector pCMV-EBNA (*Map 33*) is available (Invitrogen) if expression of the EBNA-1 gene is required.

Map 69. pRIT2T.

For high-level expression of Protein A fusion proteins.

Fusion proteins can be purified directly by affinity chromatography on IgG Sepharose columns (Protein A binds to IgG).

A portion of the λ*cro* gene has been fused with the IgG-binding domain and provides the ATG initiation codon. Genes inserted into the polylinker are expressed from the λ right promoter (λP$_R$). Translational terminators have been introduced downstream of the polylinker to improve plasmid stability.

Thermoinducible expression of the fusion protein is achieved using host strains, e.g. *E. coli* N4830-1, carrying the temperature-sensitive repressor cI857.

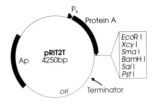

Map 70. pRSET.

A multipurpose vector for transcription of inserts using the T7 promoter or expression of proteins with an N-terminal polyhistidine tag. The polylinker is available in each of the three different reading frames.

The N-terminal histidine tag facilitates one-step purification by metal-chelate chromatography; subsequent cleavage of the purified product with enterokinase releases the native protein.

The f1 origin allows production of single-stranded DNA.

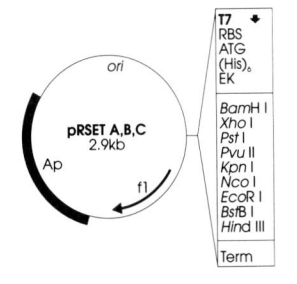

Map 71. pSE280 and related vectors.

For cloning and high-level expression of recombinant proteins in *E. coli*.

These vectors have a strong *trc* promoter, and pSE380, pSE420 and pTrcHis A, B, C also express the *lacI^q* repressor gene for regulated expression in most *E. coli* host strains.

pSE420 and pTrcHis A,B,C have translation enhancer and ribosome binding sites from bacteriophage T7 gene 10 and a short mini-cistron to provide a prokaryotic context for translation initiation.

Expression from pTrcHis A,B,C produces a fusion protein with an N-terminal fusion histidine tag for one-step metal-chelate chromatography. Native protein can be liberated by cleavage with enterokinase.

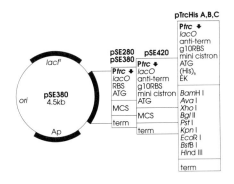

pSE280 has restriction sites for NcoI, BstEII, PflMI, PvuI, BsmI, BstBI, EcoRI, AatII, SalI, Bsu36I, SmaI, XmaI, ApaI, BamHI, BclI, BglII, BssHII, EcoRV, SphI, NheI, XbaI, KpnI, HpaI, NaeI, SnaBI, StuI, AbrII, NsiI, NotI, XmaIII, PstI, PbeI, NarI, MluI, SplI, NruI, SacII, SacI, XhoI, SpeI, SfiI, AflII, HindIII, BbvII. pSE380 has restriction sites for NcoI, PvuI, BsmI, BstBI, EcoRI, AatII, SalI, Bsu36I, SmaI, XmaI, Eco0109I, BamHI, BglII, Eco47III, NheI, XbaI, KpnI, HpaI, NaeI, SnaBI, BspMII, StuI, AbrII, NotI, XmaIII, PstI, SplI, NruI, SacII, SacI, XhoI, SpeI, BalI, SfiI, AflII, HindIII. pSE420 is identical to pSE380 except that there is a BspMI site between EcoRI and AatII.

Map 72. pSG5.

Multifunctional vector allows for the expression of cloned genes either in transfected cells, via the early SV40 promoter or *in vitro* from the T7 promoter.

The vector incorporates Intron II of the rabbit β-globin gene which facilitates splicing of the expressed transcript and a polyadenylation signal to increase expression. Greatest expression occurs in cells which also express the T antigen.

The M13 origin allows production of single-stranded DNA.

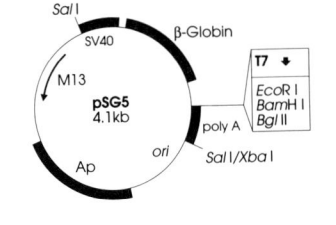

Map 73. pSL1180 and related vectors.

Vectors with exceptionally large polylinkers comprising all 64 palindromic 6 bp recognition sequences. *Not*I and *Sfi*I sites are also available.

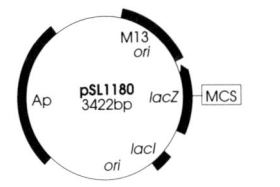

The pSL1180 polylinker has the following unique sites: *Hind*III, *Bfr*I, *Pml*I, *Sfi*I, *Spe*I, *Xho*I, *Sac*I, *Sac*II, *Nru*I, *Spl*I, *Mlu*I, *Pst*I, *Not*I, *Eag*I, *Nde*I, *Nsi*I, *Avr*II, *Stu*I, *Bsp*MII, *Sna*BI, *Hpa*I, *Kpn*I, *Xba*I, *Nhe*I, *Sph*I, *Eco*R47III, *Eco*RV, *Bss*HII, *Bgl*II, *Bcl*I, *Bam*HI, *Apa*I, *Sma*I, *Bsu*36I, *Sal*I, *Nco*I, *Bst*BI, *Eco*RI. pSL1190 has the same polylinker but in the opposite orientation.

Map 74. pSP6/T3 and related vectors.

Standard cloning vectors suitable for *in vitro* transcription from either strand using the promoters flanking the poly-linkers.

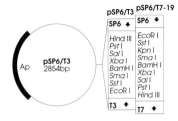

Map 75. pSP64 and related vectors.

Used as standard cloning vectors and for *in vitro* transcription (SP6 promoters).

Plasmids pSP64, pSP65 (3005 bp) and pSP64 polyA (3033 bp) are nearly identical except for the orientation and complexity of the multiple cloning site.

See also pSP70, pSP71, pSP72 and pSP73 (*Map 76*).

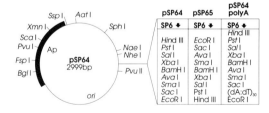

Vector Maps

Map 76. pSP70 and related vectors.

Used as standard cloning vectors and for *in vitro* transcription (SP6 promoters).

Similar to pSP64 and related vectors but with additional T7 promoter.

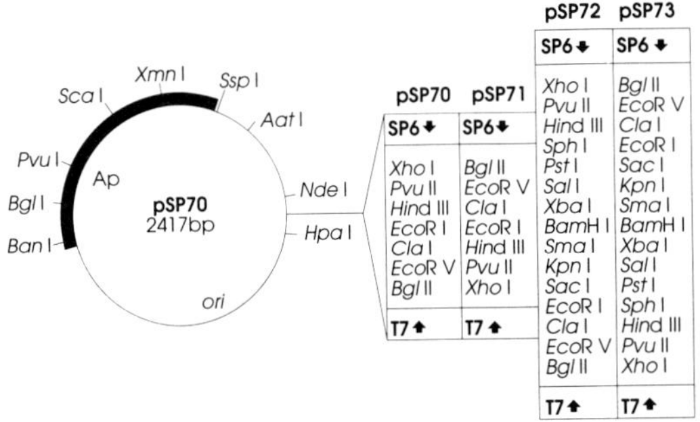

	pSP72	pSP73
	SP6 ↓	SP6 ↓
	Xho I	Bgl II
	Pvu II	EcoR V
	Hind III	Cla I
	Sph I	EcoR I
	Pst I	Sac I
	Sal I	Kpn I
	Xba I	Sma I
	BamH I	BamH I
	Sma I	Xba I
	Kpn I	Sal I
	Sac I	Pst I
	EcoR I	Sph I
	Cla I	Hind III
	EcoR V	Pvu II
	Bgl II	Xho I
	T7 ↓	T7 ↓

pSP70	pSP71
SP6 ↓	SP6 ↓
Xho I	Bgl II
Pvu II	EcoR V
Hind III	Cla I
EcoR I	EcoR I
Cla I	Hind III
EcoR V	Pvu II
Bgl II	Xho I
T7 ↓	T7 ↓

Map 77. pSPORT1.

A multifunctional vector for cloning and *in vitro* transcription from either strand using SP6 or T7 promoters.

pSPORT1 has the *lacI* and *lacZ* genes which ensures that expression of the insert DNA is repressed in the absence of IPTG. The vector features blue/white selection of recombinants and production of single-stranded DNA.

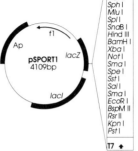

SP6 ↓
Aat II
Sph I
Mlu I
Spl I
SnaB I
Hind III
BamH I
Xba I
Not I
Sma I
Spe I
Sst I
Sal I
Sma I
EcoR I
BspM II
Rsr II
Kpn I
Pst I
T7 ↑

Vector Maps

Map 78. pSPT18 and related vectors.

Used as standard cloning vectors and for *in vitro* transcription (SP6 or T7 promoters).

Plasmids pSPT18/19 (3104 bp), pSPTBM20 (3140 bp) and pSPTBM21 (3143 bp) are nearly identical except for the orientation and complexity of the multiple cloning site.

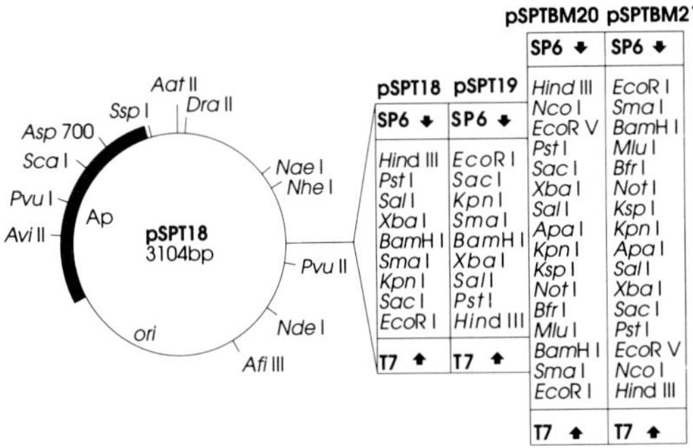

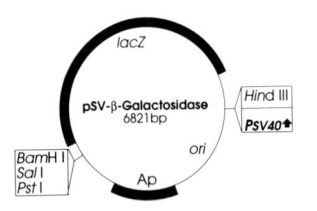

Map 79. pSV-β-Galactosidase.

Used as a positive control for monitoring transfection efficiencies in mammalian cells.

Transcription of the *lacZ* gene is driven from the SV40 early promoter and β-galactosidase activity may be measured directly in cell extracts.

Map 80. pSV2Neo.

Helper plasmid for the co-transfection of λDR2 (*Map 103*) into human cell lines.

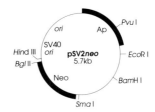

Map 81. pSVK3.

Multifunctional shuttle vector for *in vivo* expression of cloned genes in *E. coli* and mammalian cell lines.

Maintenance and expression in mammalian cells is ensured by the origin of replication, early promoter, mRNA splice site and polyadenylation signals from SV40. The ColE1 origin of replication allows maintenance in *E. coli*.

The vector may also be used for *in vitro* transcription of cloned fragments (T7 promoter) or for the production of single-stranded DNA.

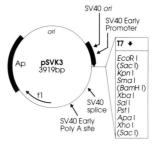

95

Vector Maps

Map 82. pSVL.

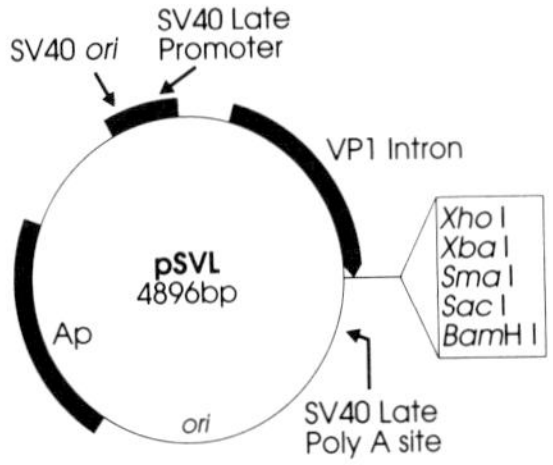

Designed for high-level transient expression in eukaryotic cells.

Genes inserted into the polylinker are expressed from the first ATG codon using the SV40 late promoter. Transcripts are spliced and polyadenylated using the SV40 VP1 processing signal.

Highest levels of expression are found in T-antigen-producing cells, e.g. Cos cells. The SV40 promoters also give high levels of expression in rodent and primate cell lines.

This shuttle vector, which contains an ampicillin-resistance gene, may be used with any suitable restriction-minus *E. coli* host strain.

Map 83. pT3T7*lac*.

Used as a standard cloning vector and for *in vitro* transcription (T7 or T3 promoters).

The *lacZ* gene allows blue/white screening of recombinants.

Map 84. pT3/T7*luc*.

pT3/T7*luc* has the 1.9 kb *luc* gene encoding firefly luciferase controlled by the *lac* promoter.

The *luc* gene can be excised by *Bam*HI, *Hind*III or *Sal*I digestion, or single-stranded RNA probes may be transcribed *in vitro* using the T3 and T7 promoters.

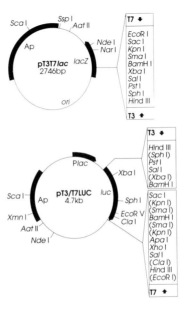

Map 85. pT7-0, pT7-1, pT7-2.

Simple vectors based on pBR322, useful for most *in vitro* transcription applications.

pT7-0 is a positive control vector used to verify that transcription is functional. pT7-1 and pT7-2 have the same polylinker regions in opposite orientations.

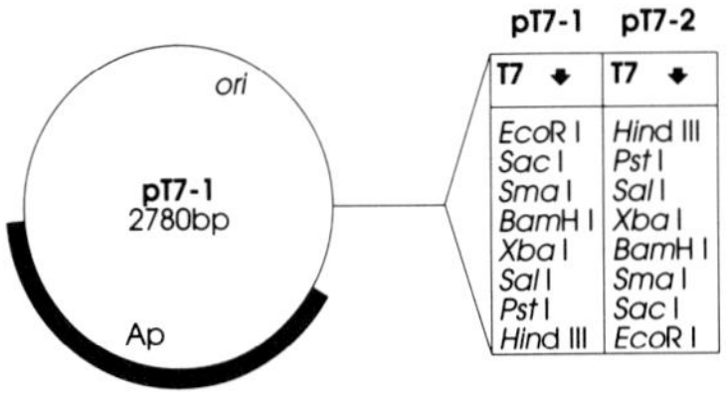

pT7-1	pT7-2
T7 ↓	T7 ↓
EcoR I	Hind III
Sac I	Pst I
Sma I	Sal I
BamH I	Xba I
Xba I	BamH I
Sal I	Sma I
Pst I	Sac I
Hind III	EcoR I

Map 86. pT7Blue and related vectors.

pT7Blue T is a general-purpose vector for cloning PCR products. Features include a T7 promoter for *in vitro* transcription and flanking *Nde*I/*Bam*HI sites for transfer of inserts to pET expression vectors (*Map 38*). A PCR cloning version of the vector is available in which the *Eco*RV site has been replaced by a T-cloning site.

Use pTOPE for T7-driven expression of inserts as stable fusion proteins. Allows construction of epitope libraries for rapid functional screening of peptide domains. When cloning PCR products it is essential that a 5′ primer is used so that the first base (N) completes the initial codon, i.e. AGA TTN, after the vector *Eco*RV site.

pCITE-T is designed for efficient *in vitro* transcription and translation of cloned inserts (see pCITE vectors, *Map 31*).

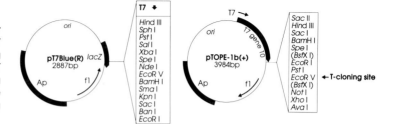

Vector Maps

Map 87. pTrc99A.

An expression vector with a strong *trc* promoter upstream, and a strong transcription termination signal downstream of the polylinker.

The *lacI^q* gene ensures adequate repression in the absence of IPTG in most *E. coli* hosts.

An *Nco*I site enables a translation start codon to be formed by ligation of insert DNA immediately downstream of the *trc* promoter. In addition, the use of *Nco*I linkers enables inserts lacking a start codon to be ligated and expressed in all three possible reading frames.

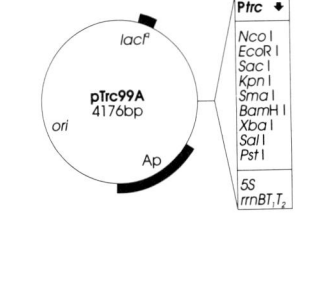

Map 88. pTRXN(+/−).

Designed for *in vitro* transcription (T7 promoter) and production of single-stranded DNA.

pTRXN− and pTRXN+ have the f1 origin in opposite orientations.

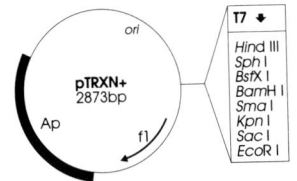

Map 89. pTZ18R and related vectors.

Multifunctional plasmids for cloning with blue/white selection of recombinants, *in vitro* transcription from T3 or T7 promoters and production of single-stranded DNA.

Vectors designated 'R' or 'U' have f1 sequences in opposite orientations.

pT7T3D is a derivative of pT7T318U which allows unidirectional cloning of cDNA as *Eco*RI/*Not*I fragments.

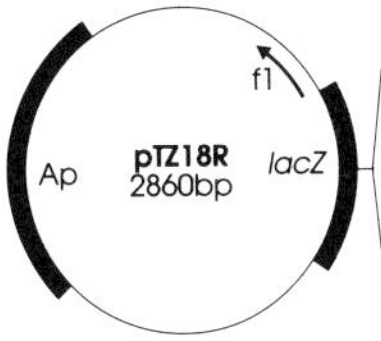

pT7/T3-18 pT7/T3-19

pTZ18R	pTZ19R	T3 ↓	T3 ↓
Hind III	EcoR I	Hind III	EcoR I
Sph I	Sac I	Sph I	Sac I
Pst I	Kpn I	Pst I	Kpn I
Sal I	Sma I	Sal I	Sma I
Xba I	BamH I	Xba I	Xba I
BamH I	Xba I	BamH I	BamH I
Sma I	Sal I	Sma I	Sal I
Kpn I	Pst I	Kpn I	Pst I
Sac I	Sph I	Sac I	Sph I
EcoR I	Hind III	EcoR I	Hind III
T7 ↑	T7 ↑	T7 ↑	T7 ↑

Map 90. pUB110.

A plasmid derived from *Staphylococcus aureus* which replicates in *Bacillus subtilis*.

There is an unique *Bgl*II site within the kanamycin resistance gene.

101

Map 91. pUC vectors.

General-purpose plasmids for cloning, sequencing and expression of foreign genes in *E. coli*.

Genes inserted into the polylinker can be expressed as fusion products under the control of the *lac* promoter. The *lacZ* gene allows blue/white screening of recombinants in appropriate host strains, e.g. JM101.

pNEB193 is a variant of pUC19 with additional sites for *Asc*I, *Pac*I and *Pme*I.

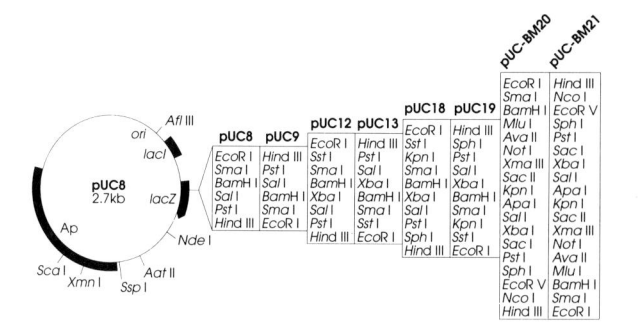

Map 92. pUR222.

DNA fragments cloned in this vector can be sequenced by the Maxam–Gilbert technique without isolation of labeled fragments.

The *lacZ* gene allows blue/white screening of recombinants.

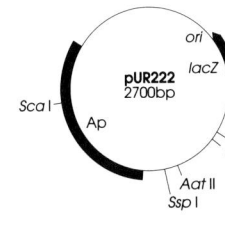

Map 93. pWE15 and related vectors.

Cosmid vectors (insert size 30–42 kb) for genomic cloning.

Inserts are flanked by T3 and T7 bacteriophage promoters for generation of end-specific transcripts.

Designed for high-resolution restriction mapping and rapid chromosomal walking.

The T3 and T7 promoters plus insert may be isolated as a *Not*I fragment.

This vector has a neomycin marker and an SV40 promoter for selection and expression in eukaryotic cells.

The neomycin gene has been replaced with the dihydro-folate reductase gene to produce the cosmid pWE16.

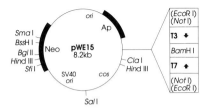

Map 94. pXPRS.

For protein expression in eukaryotic cells.

Gene expression of cloned genes is driven by an SV40 early region promoter, and there are SV40 splice sites and polyadenylation signals for correct processing of cloned transcripts.

The f1 intragenic region facilitates production of single-stranded DNA. pXPRS− is identical to pXPRS+ except for the orientation of the f1 region.

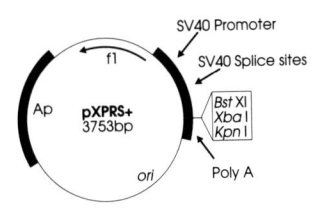

Map 95. pXT1.

For expression of exogeneous genes in stably transfected cells and transgenic mice.

The position of the LTR (from Moloney murine leukemia virus) allows highly efficient, stable transfection of the region within the LTRs. There is a selectable neomycin resistance gene.

The vector also contains the herpes simplex virus thymidine kinase (TK) promoter, which is active in embryonal cells and in a wide variety of tissues in mice.

The pBR322 sequence encodes for ampicillin and tetracycline resistance.

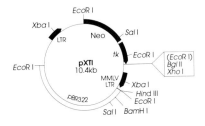

Map 96. pYAC vectors.

For cloning large DNA fragments for genome mapping and sequencing projects.

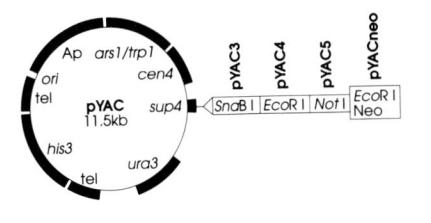

Insertion of the foreign DNA at the unique restriction sites in the *sup4* gene results in recombinants which lack expression of the ochre suppressor, i.e. colonies of recombinants are red compared to the white wild-type cells.

pYACneo has a neomycin resistance gene for selection in mammalian cells.

Map 97. pYES2 and related vectors.

High copy number episomal vector for inducible expression of recombinant proteins in *Saccharomyces cerevisiae*.

pYESHisABC allows the expression of the recombinant protein as an N-terminal fusion product with a histidine tag. This facilitates one-step purification by metal-chelate chromatography followed by subsequent cleavage with enterokinase. The MCS are also available in three different reading frames.

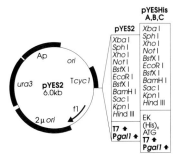

Map 98.   SuperCos.

Similar to pWE15 (*Map 93*).

SuperCos has two *cos* sites, therefore dephosphorylation of vector DNA to prevent ligation of cosmid concatamers is not necessary.

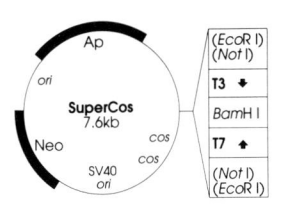

Map 99.   YEp24.

Episomal cloning vector maintained at high copy number in *S. cerevisiae*.

The plasmid can replicate in both *E. coli* (pBR322 origin of replication) and in *S. cerevisiae* (2μ plasmid sequence).

A tetracycline gene is present but separated from its promoter by the *ura3* sequence, resulting in variable antibiotic resistance.

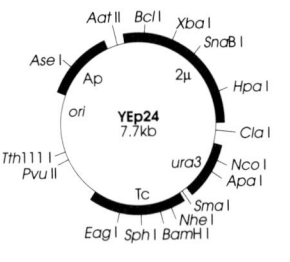

Map 100. YIp5.

A plasmid for the stable integration of constructs into the *S. cerevisiae* chromosome.

The plasmid replicates in *E. coli* but not in yeast, therefore maintenance of the recombinant DNA is dependent on successful recombination into the yeast genome.

The *ura3* gene allows selection in yeast and may sometimes be used as a region of homology for recombination into the chromosomes.

Map 101. λ2001.

DNA (9–23 kb) may be inserted into the inverted MCS which are located at either end of the *red⁺ gam⁺* stuffer fragment.

Recombinants are obtained using *spi*/P2 selection.

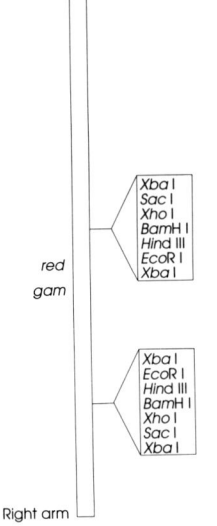

Map 102. λBlueMid.

A cDNA cloning vector (inserts up to 7 kb) easily converted to a plasmid by *Not*I digestion and religation.

The cloning site is flanked by T3 and T7 promoters for *in vitro* transcription and bidirectional restriction mapping.

An M13 origin (both + and − orientations) enables the production of single-stranded templates.

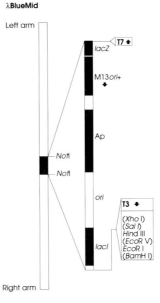

λBlueMid

Map 103. λDR2.

Lambda-based vector which contains the embedded plasmid, pDR2.

pDR2 has an RSV-LTR promoter for efficient expression of recombinant proteins in human cells. The plasmid is maintained at a copy number of approximately 10–20 per cell.

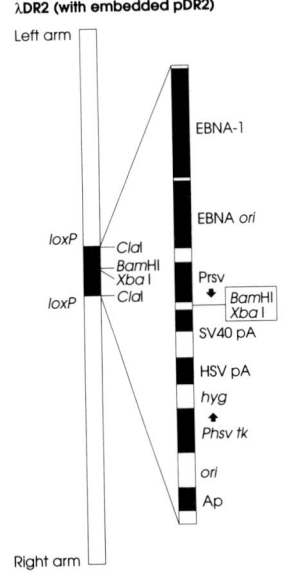

λDR2 (with embedded pDR2)

Map 104. λEMBL3 and related vectors.

Replacement vectors for genomic cloning (insert size 9–23 kb).

Double digestion of the vector prevents religation of the stuffer fragment.

Recombinants are obtained using *spi*/P2 selection.

Use of *E. coli* SRB and SRB (P2) hosts enhances stability of methylated genomic DNA sequences and some nonstandard DNA structures.

Commercially available (Clontech) genomic libraries have been constructed in a derivative of λEMBL3 with flanking SP6 and T7 promoters.

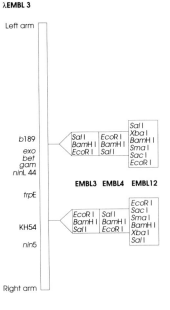

λEMBL 3

Vector Maps

Map 105. λFIX™II and λDASH™II.

Replacement vectors (insert size 9–23 kb) for genomic cloning.

Inserts are flanked by T3 and T7 bacteriophage promoters for generation of end-specific transcripts.

Designed for high-resolution restriction mapping and rapid chromosomal walking.

The T3 and T7 promoters plus insert may be isolated as a *Not*I fragment.

Recombinants are obtained using *spi*/P2 selection.

Use of *E. coli* SRB and SRB (P2) hosts enhances stability of methylated genomic DNA sequences and some nonstandard DNA structures.

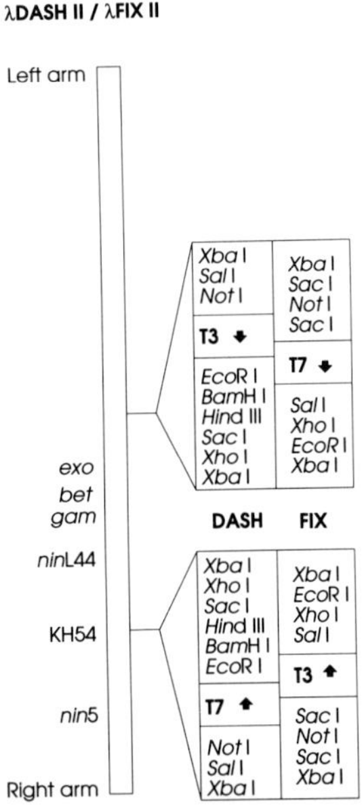

λDASH II / λFIX II

Map 106. λGEM-2 and λGEM-4.

Cloning capacity is 0.7–1 kb for λGEM-2 and 0.4–1 kb for λGEM-4.

Vectors for highly efficient directional or nondirectional cDNA cloning and selective amplification of either sense or antisense cDNA sequences using the SP6 or T7 promoters.

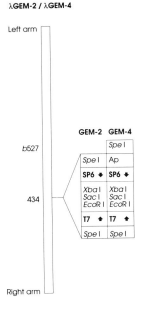

λGEM-2 / λGEM-4

Vector Maps

Map 107. λGEM-11 and λGEM-12.

Multifunctional replacement vectors for cloning genomic DNA fragments (9–23 kb).

Flanking SP6 and T7 promoters allow RNA probe synthesis, thus simplifying chromosomal walking.

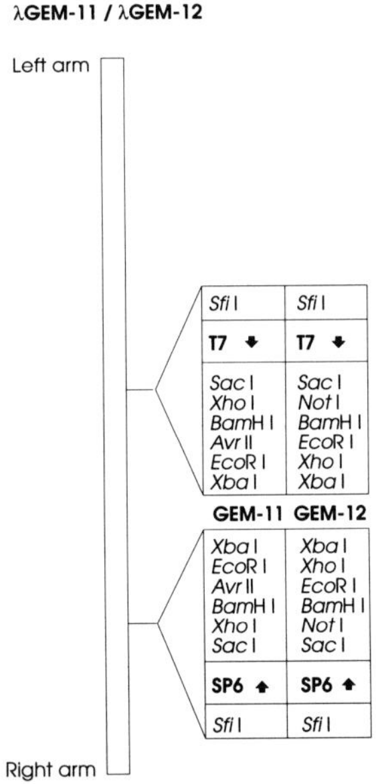

λGEM-11 / λGEM-12

	λGEM-11	λGEM-12
Left arm		
	Sfi I	*Sfi* I
	T7 ✦	**T7** ✦
	Sac I *Xho* I *Bam*H I *Avr* II *Eco*R I *Xba* I	*Sac* I *Not* I *Bam*H I *Eco*R I *Xho* I *Xba* I
	GEM-11	**GEM-12**
	Xba I *Eco*R I *Avr* II *Bam*H I *Xho* I *Sac* I	*Xba* I *Xho* I *Eco*R I *Bam*H I *Not* I *Sac* I
	SP6 ✦	**SP6** ✦
	Sfi I	*Sfi* I
Right arm		

Map 108. λgt10.

Inserts <7.6 kb can be cloned into the unique *Eco*RI site within the *cI* repressor gene.

Recombinants produce clear plaques compared to the turbid plaques of vector alone.

E. coli C600 strains provide visual identification of recombinants, whereas NM514 selects against nonrecombinant phage.

λgt10

Left arm

*b*527

*imm*434 — *Eco*R I

Right arm

Map 109. λgt11.

For construction of gene libraries (inserts <7.2 kb) which can be screened by either antibodies or nucleic acid probes.

Inserted sequences are expressed as β-galactosidase fusion proteins.

Use *E. coli* Y1090 as host for analysis and screening or Y1089 for lysogeny and overproduction of fusion protein.

Several variants of λgt11 exist for directional cloning, e.g. λgt11 *Sfi-Not* (Promega). λgt11D is a similar vector to λgt11 *Sfi-Not*.

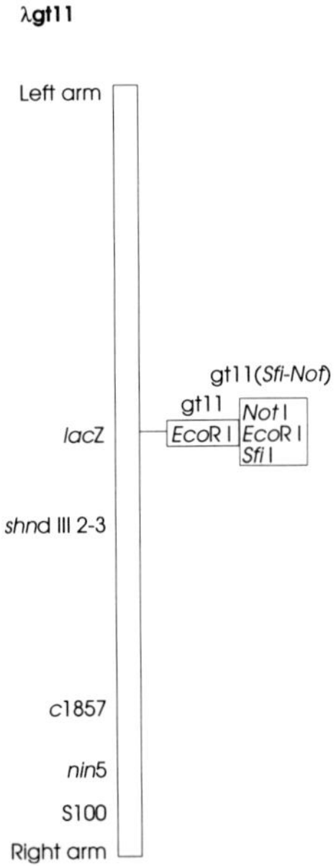

Map 110. λgt18 and related vectors.

Derivatives of λgt11 (*Map 109*) with a deletion at *ssI*λ1–2, i.e. between *Sal*I sites 1 and 2 in the genome. This allows the use of a modified cloning site with a unique *Sal*I site in addition to the *Eco*RI site of λgt11.

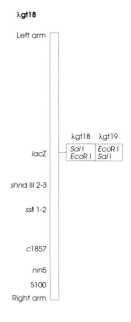

119

Map 111. λgt20 and related vectors.

Orientation-specific cloning of cDNA (< 7 kb) and expression of insert as a β-galactosidase fusion protein.

Recombinants may be screened with either nucleic acid or antibody probes.

Blue/white screening of recombinants in *E. coli lac⁻* hosts.

λgt22A has SP6 and T7 promoters flanking the cloning site.

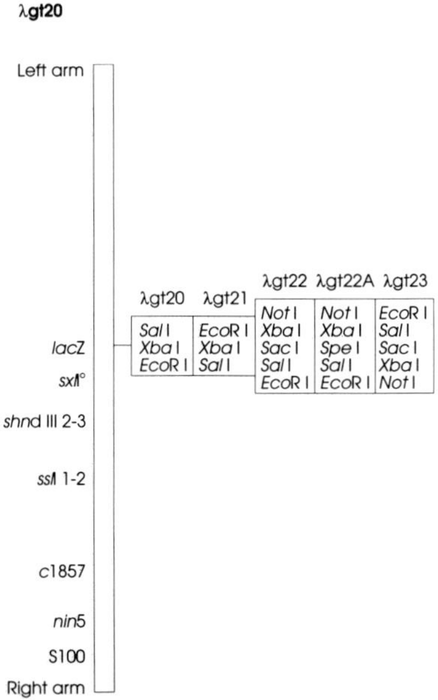

λgt20

Left arm	

	λgt20	λgt21	λgt22	λgt22A	λgt23
			Not I	*Not* I	*Eco*R I
	Sal I	*Eco*R I	*Xba* I	*Xba* I	*Sal* I
lacZ	*Xba* I	*Xba* I	*Sac* I	*Spe* I	*Sac* I
*sxt*ᴾ	*Eco*R I	*Sal* I	*Sal* I	*Sal* I	*Xba* I
			*Eco*R I	*Eco*R I	*Not* I

shnd III 2-3	
ssl 1-2	
c1857	
nin5	
S100	
Right arm	

Map 112. λgtWES.λb.

Suitable for cloning DNA inserts of 2–17 kb using *Eco*RI or *Sac*I.

A *supF E. coli* host (LE392) should be used for propagation.

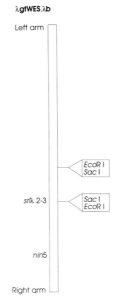

λ**gtWES.λb**

Left arm

EcoR I
Sac I

srλ 2-3

Sac I
EcoR I

nin5

Right arm

Vector Maps

Map 113. λMAX1.

For directional cloning of cDNA (up to 8 kb) and subsequent expression in yeast. Subcloning is unnecessary as an embedded form of the pYEUra3 plasmid may be rescued with helper phage.

pYEUra3 carries the Ura3 marker for selection, CEN4 and ARS1 for propagation, and the GAL1 and GAL10 promoters for expression in yeast cells.

Insert DNA will be expressed as a fusion protein from the *Eco*RI site but will require an ATG codon for expression from the *Xba*I or *Xho*I sites.

pYEUra3 also contains an origin of replication for maintenance in *E. coli* and an ampicillin gene for selection.

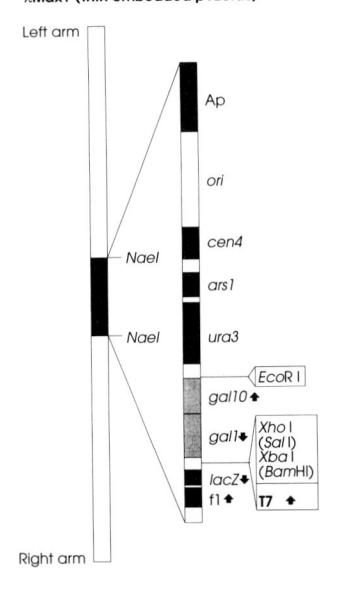

λMax1 (with embedded pYEUra3)

Map 114. λMOSSlox and λMOSElox.

Insert DNA < 8 kb.

Recombinants may be converted directly into plasmids *in vivo*. Complete plasmids, which are built into the phage, are excised by infection of hosts expressing the P1 recombinase.

λMOSSlox is for screening cDNA libraries with nucleic acid probes whereas λMOSElox is designed for expression of fusion proteins, which may be screened using antibodies. Expression of cDNA is under the control of a T7 promoter. Gene products are expressed as bacteriophage T7 gene10 fusion proteins following induction with IPTG. Expressing host strains should carry the T7 RNA polymerase gene.

λSHlox-1 (generating pSHlox-1) and λEXlox (+) are synonyms for λMOSSlox and λMOSElox.

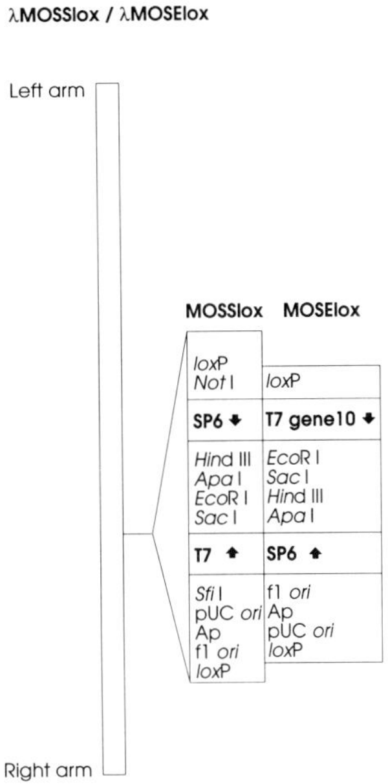

λMOSSlox / λMOSElox

Left arm

MOSSlox   MOSElox

MOSSlox	MOSElox
loxP *Not* I	*loxP*
SP6 ✦	T7 gene10 ✦
Hind III *Apa* I *EcoR* I *Sac* I	*EcoR* I *Sac* I *Hind* III *Apa* I
T7 ✦	SP6 ✦
Sfi I pUC *ori* Ap f1 *ori* *loxP*	f1 *ori* Ap pUC *ori* *loxP*

Right arm

Vector Maps

Map 115. λPOP6.

pPOP6 is a mammalian episomal expression plasmid vector embedded in λPOP6.

Insertion of cDNA into the polylinker of λPOP6 disrupts the *cI* repressor and permits plaque formation, so facilitating selection of recombinants. Plasmid is released from λPOP6 using the *cre/lox* recombinase system of bacteriophage P1.

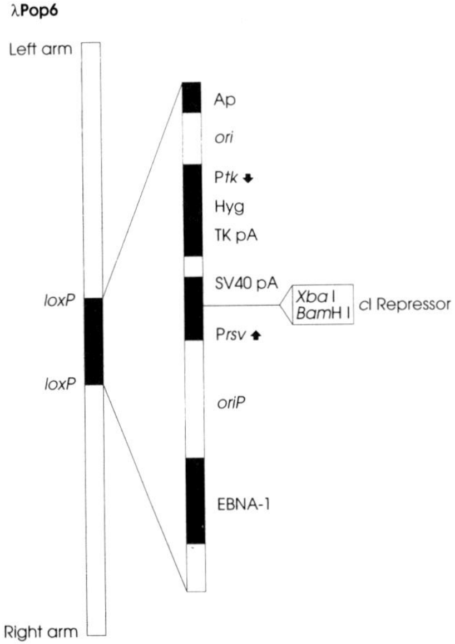

λ**Pop6**

Map 116. λYES.

For cloning of cDNA (up to 8 kb) into *Eco*RI or *Xho*I sites and subsequent expression of fusion protein in *E. coli* or yeast.

An embedded form of the pSE937 plasmid may be released (*lox* sites) and used for complementation studies in either host.

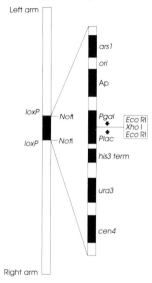

λYES (with embedded pSE937)

125

Map 117. λZAP II.

Insertion vector for construction of cDNA libraries (insert size < 10 kb) with screening using nucleic acid probes or antibodies.

Expression of inserts as *lacZ* fusion proteins.

Generation of strand-specific RNA transcripts from flanking T3 and T7 promoters.

Excision and recircularization *in vitro* of recombinant λZAP II with helper phage automatically generates subclones in pBluescript SK.

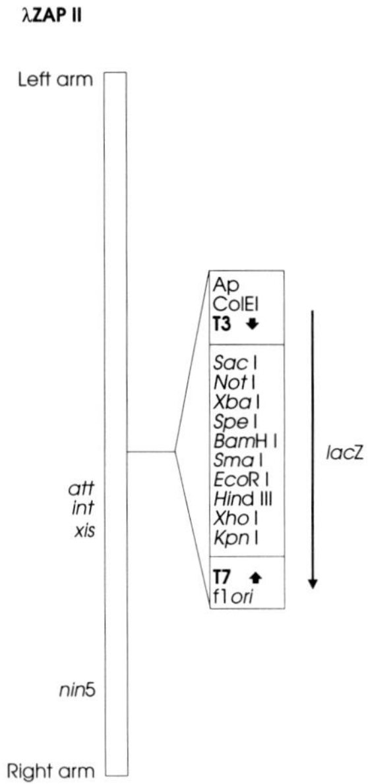

λ**ZAP II**

Chapter 4 HOST CELL INFORMATION

Table 1. *E. coli* host strains

Plasmid DNA preparation (e.g. *endA, recA*)
  AG1, AM1, DH1, DH5, DH5α, DH5αF', DH5αF'IQ, DH5αMCR, DH21, DH10B, INVαF', JM109, NovaBlue, SCS1, TOP10, XL1-Blue, XL1-Blue MR

Genomic library construction (e.g. *mcrA, mcrCB, mrr, hsdR, recA*)
  DH5αMCR, DH10B, KW251, NM554, PLK-A, PR700, PR745, XL1-Blue MRF', SURE™

Single-stranded DNA rescue (F')
  AM1, BB4, BL26 (Blue), BM28.8, BMH71-18 mutS, CJ236, DH5αF', DH5αF'IQ, DH21, DH10B, ER2267, INVαF', JM101, JM103, JM105, JM107, JM109, JM110, KS1000, MV1190, NM522, NovaBlue, PLK-F', SDM2, SRB, SURE™,TOP10F', XL1-Blue, XL1-Blue MRF'

Cloning unstable DNA structures (e.g. *recB, recJ, sbcC, umuC, uvrC*)
  SRB, SURE™

Overexpression of LacI repressor protein (*lacI*^q^)
  AM1, BB4, BL26 (Blue), BM28.8, BMH71-18 mutS, D1210, DH5αF'IQ, DH21, DH1210HP, ER2267, JM101, JM103, JM105, JM107, JM109, JM110, KS1000, MV1190, NM522, NovaBlue, PLK-F', SDM2, SRB, SURE™, XL1-Blue, XL1-Blue MRF'

Continued

Table 1. *E. coli* host strains, *continued*

Blue/white colour selection (*lacZΔM15*)
  BB4, BL26 (Blue), BM28.8, BMH71-18 mutS, DH5α, DH5αF′, DH5αF′IQ, DH5αMCR, DH10B, ER2267, INVαF′, JM101, JM103, JM105, JM107, JM109, JM110, MV1190, NM522, NovaBlue, PLK-F′, SDM2, SRB, SURE™, TB1, TOP10, XL1-Blue, XL1-Blue MRF′

Lack of methylation at GATC, CCAGG or CCCTGG sites (*dam, dcm*)
  B834; B834 (DE3), B834 (DE3)pLysS (*dcm*); BL21, BL21 (DE3), BL21 (DE3)pLysS, BL21 (DE3)pLysE (*dcm*); BL26, BL26Blue (*dcm*); BLR, BLR (DE3), BLR (DE3)pLysS (*dcm*); GM2163 (*dam, dcm*), JM110 (*dam, dcm*)

Homologous-recombination-deficient strains (e.g. *recA*1)
  AG1, AM1, BW313/P3, CES201, D1210, DH1, DH5, DH5α, DH5αF′, DH5αF′IQ, DH5αMCR, DH21, DH10B, DH1210HP, ER2267, HB101, HMS174, HMS174 (DE3), HMS174 (DE3)pLysS, HMS174 (DE3)pLysE, INVαF′, JM109, JM109 (DE3), NM554, NovaBlue, NovaBlue (DE3), PLK-A, SCS1, TOP10, TOP10F′, XL1-Blue, XL1-Blue MRF′, XL1-Blue MR

Expression from T7 promoters
  B834 (DE3), B834 (DE3)pLysS, BL21 (DE3), BL21 (DE3)pLysS, BL21 (DE3)pLysE, BLR (DE3), BLR (DE3)pLysS, HMS173 (DE3), HMS174 (DE3)pLysS, HMS174 (DE3)pLysE, JM109 (DE3), NovaBlue (DE3)

Carrier of the P1 prophage
  BM28.8, JM103

Carrier of the P2 prophage
  NM539, P2392, SRB(P2)

Carrier of the P3 plasmid
  BW313/P3, CJ236/P3, MC1061/P3, XS127/P3

Kunkel method of site-directed mutagenesis (*dut-1, ung-1*)
  CJ236

pET vector host strains
  B834, B834 (DE3), B834 (DE3)pLysS, BL21, BL21 (DE3), BL21 (DE3)pLysS, BL21 (DE3)pLysE, BLR, BLR (DE3), BLR (DE3)pLysS, HMS174,
  HMS174 (DE3), HMS174 (DE3)pLysS, HMS174 (DE3)pLysE

General λ propagation
  K802, LE392, NM538

Hosts for λgt10
  C600, C600(*hfl*)

Hosts for λgt11
  Y1088, Y1089, Y1089r−, Y1090r−

Hosts for *spi*⁻ selection and growth
  CES200, CES201, NM539, P2392

Hosts for λORF8
  MC1061

Host Cell Information

Table 2. Genotypes of *E. coli* host strains

Strain	Genotype	Supplier
AG1	F$^-$, *recA1*, *endA1*, *gyrA96*, *thi−1*, *hsdR17* (r_K^-, m_K^+), *supE44*, *relA1*	Str
AM1	*supE44*, *hsdR17*, *recA1*, *endA1*, *gyrA96*, *thi-1*, *relA1*, (pItsCRE3), F'[*lacI^q*, *lacZ$^+$*, *lacAB$^+$*]	Inv
B834	F$^-$, *ompT*, *hsdS* (r_B^-, m_B^-), *gal*, *dcm*, *met*	Nov
B834(DE3)	F$^-$, *ompT*, *hsdS* (r_B^-, m_B^-), *gal*, *dcm*, *met* (DE3)	Nov
B834 (DE3)pLysS	F$^-$, *ompT*, *hsdS* (r_B^-, m_B^-), *gal*, *dcm*, *met* (DE3) pLysS (Cm^r)	Nov
BB4	*e14$^-$(mcrA)*, *hsdR514*, *supE44*, *supF58*, *lacY1* or *Δ(lacIZY)6*, *galK2*, *galT22*, *metB1*, *trpR55*, *Δ(argF-lac) U169*, [F', *lacI^{q}ZΔM15*, *proAB*, Tn10 (Tc^r)]	Str
BL21	F$^-$, *ompT*, *hsdS* ($r_B^- m_B^-$), *gal*, *dcm*	Nov
BL21 (DE3)	F$^-$, *ompT*, *hsdS* ($r_B^- m_B^-$), *gal*, *dcm* (DE3)	Nov
BL21 (DE3)pLysS	F$^-$, *ompT*, *hsdS* ($r_B^- m_B^-$), *gal*, *dcm* (DE3), pLysS (Cm^r)	Nov
BL21 (DE3)pLysE	F$^-$, *ompT*, *hsdS* ($r_B^- m_B^-$), *gal*, *dcm* (DE3), pLysE (Cm^r)	Nov
BL26	F$^-$, *ompT*, *hsdS* ($r_B^- m_B^-$), *gal*, *dcm*, *lac*	Nov
BL26Blue	*ompT*, *hsdS* ($r_B^- m_B^-$), *gal*, *dcm*, *lac* [F', *proAB*, *lacI^{q}ZΔM15*::Tn10 (Tc^r)]	Nov
BLR	F$^-$, *ompT*, *hsdS* (r_B^-, m_B^-), *gal*, *dcm*, *Δ(srl-recA)306*::Tn10 (Tc^r)	Nov
BLR (DE3)	F$^-$, *ompT*, *hsdS* (r_B^-, m_B^-), *gal*, *dcm*, *Δ(srl-recA)306*::Tn10 (Tc^r) (DE3)	Nov
BLR (DE3)pLysS	F$^-$, *ompT*, *hsdS* (r_B^-, m_B^-), *gal*, *dcm*, *Δ(srl-recA)306*::Tn10 (Tc^r) (DE3) pLysS (Cm^r)	Nov
BM28.8	*supE*, *thi*, *Δ(lac-proAB)* λ*imm^{434}* (Km^r), P1 (Cm^r), *hsdR* (r_{K12}^-, m_{K12}^+) [F', *traD36*, *proAB$^+$*, *lacI^{q}ZΔM15*]	Nov
BMH 71−18 *mutS*	*thi-1*, *supE*, *Δ(lac-proAB)*, [*mutS*::Tn10], [F', *proAB*, *laqI^{q}ZΔM15*]	Pro, Clo
BW313/P3	F$^-$, *hfr*, *lysA*, *dut*, *ung*, *thi-1*, *recA*, *spoT1* [P3: amber Ap^r, amber Tc^r, Km^r]	Inv
C600	F$^-$, *hsdR*, *thi-1*, *thr-1*, *leuB6*, *lacY1*, *tonA21*, *supE44*	Inv, Clo, Str
C600*hfl*	F$^-$, *hsdR*, *thi-1*, *thr-1*, *leuB6*, *lacY1*, *tonA21*, *supE44*, *hflA150*, [*chr*::Tn10]	Inv, Clo
CAG597	F$^-$, *rpoH165(am)*, *zhg*::Tn10, *lacZ(am)*, *trp(am)*, *pho(am)*, *supC(ts)*, *mal(am)*, *rpsL*	Bio

CAG626	F⁻, *lon, lacZ(am), trp(am), pho(am), supC(ts), mal(am), rpsL*	Bio
CAG629	F⁻, *rpoH165(am), zhg::Tn10, lacZ(am), trp(am), pho(am), supC(ts), mal(am), rpsL, lon*	Bio
CAG748	F⁻, *thi, leu, lacY, tonA, supE44, Δ(lac)X90, dnaJ259, thr::Tn10* (Tcʳ)	Bio
CES200	F⁻, *hsdR, recBC, sbcBC*	Bio
CES201	F⁻, *recA, sbcB15, recB21, recC22, hsdR*	Inv
CJ236	*dut-1, ung1, thi-1, relA1*, [pCJ105 (Cmʳ) F']	Inv
CJ263/P3	*dut-1, ung-1, thi-1, relA1*, [pCJ105 Cmʳ F'] [P3: *amber* Apʳ, *amber* Tcʳ, Kmʳ]	Inv
D1210	F⁻, *hsdS20* (r_B⁻, m_B⁻), *supE44, ara14, galK2, lacY1, proA2, rspL20, xyl-5, mtl-1, recA13, mcrB, mrr, lacIq, lacY⁺*	Str
D1210HP	F⁻, *thi-1, hsdS20* (r_B⁻, m_B⁻), *supE44, recA13, ara-14, leuB6, proA2, lacY1, rpsL20* (Strʳ), *xyl-5, mtl-1, lacIq, lacY⁺, λxis⁻kil⁻cI857*	Str
DH1	F⁻, *recA1, endA1, gyrA96, thi-1, hsdR17* (r_K⁻, m_K⁺), *supE44, relA1*	Inv
DH5	F⁻, *recA1, endA1, hsdR17* (r_K⁻, m_K⁺), *supE44, thi-1, gyrA96, relA1*	BRL, USB
DH5α	F⁻, *φ80dlacZΔM15, recA1, endA1, gyrA96, thi-1, hsdR17* (r_K⁻, m_K⁺), *supE44, relA1, deoR, Δ(lacZYA−argF)U169*	NBS, BRL, Clo
DH5αF	F', *φ80dlacZΔM15, Δ(lacZYA-argF)U169, deoR, recA1, endA1, hsdR17* (r_K⁻, m_K⁺), *supE44, λ⁻, thi-1, gyrA96, relA1*	BRL
DH5αF'IQ	F', *φ80dlacZΔM15, Δ(lacZYA−argF)U169, deoR, recA1, endA1, hsdR17* (r_K⁻, m_K⁺), *supE44, λ⁻, thi-1, gyrA96, relA1* [F', *proAB⁺, lacIqZΔM15, zzf::Tn5* (Kmʳ)]	BRL
DH5αMCR	F⁻, *mcrA, (mrr-hsd, RMS-mcrBC), φ80dlacZΔM15, Δ(lacZYA−argF)U169, deoR, recA1, endA1, supE44, λ⁻, thi-1, gyrA96, relA1*	BRL
DH21	*recA1, endA1, gyrA96, thi-1, hsdR17* (r_K⁻, m_K⁺), *supE44, relA1*, [F', *lacIq, lacZ⁺, proAB⁺*]	Inv
DH10B	F', *mcrA, Δ(mrr-hsd, RMS-mcrBC), φ80dlacZΔM15, ΔlacX74, deoR, recA1, φaraD139, Δ(ara,leu)7697, galU, galK, λ⁻, rpsL, endA1, nupG*	BRL
ED8739	F⁻, *metB, supE, supF, hsdS* (r_K12⁻, m_K12⁻)	Nov

Continued

Host Cell Information

Table 2. Genotypes of *E. coli* host strains, *continued*

Strain	Genotype	Supplier
ER1370	F$^-$, *fhuA2*, Δ*(lacZ)r1*, *supE44*, *trp31*, *his-1*, *argG6*, *rpsL104*(Str^r), *xyl-7*, *mtl-2*, *metB1*, *serB28*	Bio
ER1378	F$^-$, *fhuA2*, Δ*(lacZ)r1*, *supE44*, *trp31*, *his-1*, *argG6*, *rpsL104*(Str^r), *xyl-7*, *mtl-2*, *metB1*, *serB28*, *mcrB1*, *hsdR2* (r$_K^-$, m$_K^+$), Ser$^+$	Bio
ER1381	F$^-$, *fhuA2*, Δ*(lacZ)r1*, *supE44*, *trp31*, *his-1*, *argG6*, *rpsL104*(Str^r), *xyl-7*, *mtl-2*, *metB1*, *serB28*, *hsdR2* (r$_K^-$, m$_K^+$), Ser$^+$	Bio
ER1398	F$^-$, *endA1*, *hsdR17* (r$_K^-$, m$_K^+$), *supE44*, *thi-1*, *relA1?*, *rfbD1?*, *spoT1?*, *mcrB1*, *hsdR2* (r$_K^-$, m$_K^+$)	Bio
ER1414	F$^-$, *IN(rrnD-rrnE)1*, *mcrB1*, *hsdR2* (r$_K^-$, m$_K^+$), *zjj202::Tn10* (Tc^r), *serB28*	Bio
ER1458	F', Δ*(lacU169)*, *lon-100*, *hsdR*, *araD139*, *rpsL*(Str^r), *supF*, *mcrA*, *trp$^+$*, *zjj202::Tn10* (Tc^r), *hsdR2* (r$_K^-$, m$_K^+$), *mcrB1*	Bio
ER1562	F$^-$, *endA1*, *hsdR17* (r$_K^-$, m$_K^+$), *supE44*, *thi-1*, *relA1?*, *rfbD1?*, *spoT1?*, *mcrA1272::Tn10* (Tc^r), *hsdR2* (r$_K^-$, m$_K^+$), *mcrB1*	Bio
ER1563	F$^-$, *endA1*, *hsdR17* (r$_K^-$, m$_K^+$), *supE44*, *thi-1*, *relA1?*, *rfbD1?*, *spoT1?*, *mcrA1272::Tn10* (Tc^r)	Bio
ER1564	F$^-$, *fhuA2*, Δ*(lacZ)r1*, *supE44*, *trp31*, *his-1*, *argG6*, *rpsL104*(Str^r), *xyl-7*, *mtl-2*, *metB1*, *serB28*, *mcrA1272::Tn10* (Tc^r), *hsdR2* (r$_K^-$, m$_K^+$), Ser$^+$	Bio
ER1565	F$^-$, *fhuA2*, Δ*(lacZ)r1*, *supE44*, *trp31*, *his-1*, *argG6*, *rpsL104*(Str^r), *xyl-7*, *mtl-2*, *metB1*, *serB28*, *mcrA1272::Tn10* (Tc^r), *hsdR2* (r$_K^-$, m$_K^+$), Ser$^+$, *mcrB1*	Bio
ER1647	F$^-$, *fhuA2*, Δ*(lacZ)r1*, *supE44*, *trp31*, *mcrA1272::Tn10* (Tc^r), *his-1*, *rpsL104* (Str^r), *xyl-7*, *mtl-2*, *metB1*, Δ*(mcrC-mrr)102::Tn10* (Tc^r), *recD1014*	Bio, Nov
ER1648	F$^-$, *fhuA2*, Δ*(lacZ)r1*, *supE44*, *trp31*, *mcrA1272::Tn10* (Tc^r), *his-1*, *rpsL104* (Str^r), *xyl-7*, *mtl-2*, *metB1*, Δ*(mcrC-mrr)102::Tn10* (Tc^r)	Bio
ER1821	F$^-$, *e14$^-$(mcrA$^-$)*, *endA1*, *supE44*, *thi-1*, *relA1?*, *rfbD1?*, *spoT1?*, Δ*(mcrC-mrr)114::IS10*	Bio
ER2267	*e14$^-$(mcrA$^-$)*, *endA1*, *supE44*, *thi-1*, *relA1?*, *rfbD1?*, *spoT1?*, Δ*(mcrC-mrr)114::IS10*, Δ*(argF-lac)U169*, *recA1* [F', *proAB$^+$*, *lacI^q*Δ*(lacZ)M15*, *zzf::mini-Tn10* (Km^r)]	Bio

GM2163	F⁻, *ara-14, leuB6, thi-1, fhuA31, lacY1, tsx-78, galK2, galT22, supE44, hisG4, rpsL136* (Str^r), *xyl-5, mtl-1, dam13::Tn9* (Cm^r), *dcm-6, mcrB1, hsdR2* (r$_K^-$, m$_K^+$), *mcrA*	Bio
HB101	F⁻, *hsdS20* (r$_B^-$, m$_B^-$), *supE44, recA13, ara-14, galK2, proA2, lacY1, rpsL20* (Str^r), *xyl-5, mtl-1*	Inv, USB, Pro, BRL, NBS
HMS174	F⁻, *recA, hsdR* (r$_{K12}^-$, m$_{K12}^+$), Rif^r	Nov
HMS174 (DE3)	F⁻, *recA, hsdR* (r$_{K12}^-$, m$_{K12}^+$), Rif^r (DE3)	Nov
HMS174 (DE3)pLysS	F⁻, *recA, hsdR* (r$_{K12}^-$, m$_{K12}^+$), Rif^r (DE3), pLysS, Cm^r	Nov
HMS174 (DE3)pLysE	F⁻, *recA, hsdR* (r$_{K12}^-$, m$_{K12}^+$), Rif^r (DE3), pLysE, Cm^r	Nov
INVαF′	F′, *endA1, recA1, hsdR17* (r$_K^-$, m$_K^+$), λ⁻, *supE44, thi-1, gyrA96, relA1,φ80ΔlacΔM15, Δ(lacZYA-argF)U169, deoR*	Inv
JM101	*supE, thi-1, Δ(lac-proAB)*[F′, *traD36, proAB, lacI^{q}ZΔM15*]	Inv, Clo, USB, Str, NBS, Bio
JM103	*endA1, supE, sbcBC, thi-1, rpsL*(Str^r), *Δ(lac-pro)*, (P1), (r$_K^+$, m$_K^+$) (r$_{P1}^+$, m$_{P1}^+$) [F′, *traD36, lacI^{q}Δ(lacZ)M15, proAB$^+$*]	Bio
JM105	*endA1, thi-1, rpsL*(Str^r), *sbcB15, hsdR4* (r$_K^-$, m$_K^+$), *Δ(lac-proAB)*, [F′, *traD36, proAB, lacI^{q}ZΔM15*]	Pha, Bio
JM107	*e14⁻, (mcrA⁻), Δ(lac-proAB), thi-1, gyrA96*(Nal^r), *endA1, hsdR17* (r$_K^-$, m$_K^+$), *relA1, supE44* [F′, *traD36, lacI^{q}Δ(lacZ)M15, proAB$^+$*]	Bio
JM109	*endA1, recA1, gyrA96, thi-1, hsdR17* (r$_K^-$, m$_K^+$), *relA1, supE44, Δ(lac-proAB)* [F′, *traD36, proAB, lacI^{q}ZΔM15*]	Inv, Clo, USB, Str, Pro NBS, Bio
JM109 (DE3)	*endA1, recA1, gyrA96, thi-1, hsdR17* (r$_K^-$, m$_K^+$), *relA1, supE44, Δ(lac-proAB)* [F′, *traD36, proAB, lacI^{q}ZΔM15*], λ(DE3)	Pro
JM110	*rpsL, thr, thi-1, leu, hsdR17* (r$_K^-$, m$_K^+$), *lacY, galK, galT, ara, tonA, tsx, dam, dcm, supE44, Δ(lac-proAB)* [F′, *traD36, proAB, lacI^{q}ZΔM15*]	*Inv, Str*
K802	F⁻, *mcrB1, hsdR2* (r$_K^-$, m$_K^+$), *galK2, galT22, supE44, metB1*	Clo, Str

Continued

Host Cell Information

Table 2. Genotypes of *E. coli* host strains, *continued*

Strain	Genotype	Supplier
KS1000	[F', *lacI^q*, *lac$^+$pro$^+$*] *ara*, Δ(lac-pro), *nalA*, *argI(am)*, *rif*, *thi-1*, Δ(tsp)::K$_m$^{r}, *eda-51*::Tn10(Tc^r)	Bio
KW251	F$^-$, *supE44*, *galK2*, *galT22*, *metB1*, *hsdR2*, *mcrB1*, *mcrA*, [argA81::Tn10], *recD1014*	Pro
LE392	F$^-$, *hsdR574* (r$_K$$^-$, m$_K$$^+$), *supE44*, *supF58*, *lacY1* or Δ(lacIZY)6, *galK2*, *galT22*, *metB1*, *trpR55*	Str, Pro, Nov
MC1061	F$^-$, *araD139*, Δ(ara-leu)7679, Δ(lac)X74, *thi-1*, *galU*, *galK*, *hsdR2* (r$_K$$^-$,m$_K$$^+$), *mcrB1*, *rpsL*(Str^r)	Clo, USB
MC1061/P3	F$^-$, *araD139*, Δ(ara-leu)7679, Δ(lac)X74, *thi-1*, *galU*, *galK*, *hsdR2* (r$_K$$^-$,m$_K$$^+$), *mcrB1*, *rpsL*(Str^r) [P3: amber Ap^r, amber Tc^r, Km^r]	Inv, Clo
MM294	F$^-$, *endA1*, *hsdR17* (r$_K$$^-$, m$_K$$^+$), *supE44*, *thi-1*	Clo
MV1190	*supE*, Δ(srl-recA)306::Tn10 Δ(lac-pro), *thi* [F', *traD36*, *proAB$^+$*, *lacI^{q}lacZΔM15*]	USB
N99cI$^+$	*galK*, *strA*, λI$^+$	Pha
N4830−1	F$^-$, *galK8*, *thi1*, *thr1*, *leuB6*, *lacY1*, *fhuA21*, *supE44*, *rfbD1*, *mcrA1*, *his*, *ilv*, Δ(hemF-esp), Δ(bio-uvrB), Δ[λ,Bam, Δ(cro-attR), N$^+$], *cI857*	Pha
NM514	F$^-$, *hsdR514* (r$_K$$^-$ m$_K$$^-$), *argH*, *galE*, *galX*, *lycB7*, *strA*, (hfl$^+$)	Str
NM522	*supE*, *thi*, Δ(lac-proAB), Δhsd5 (r$^-$, m$^-$) [F', *proAB*, *lacI^{q}ZΔM15*]	Inv, Str, Pha, Pro
NM554	F$^-$, *recA13*, *araD139*, Δ(ara-leu)7696, Δ(lac)I7A, *galU*, *galK*, *hsdR*, *rpsL*(Str^r), *mcrA*, *mcrB*	Str
NM538	F$^-$, *supF*, *hsdR* (r$_K$$^-$, m$_K$$^+$), *trpR*, *lacY*	Pro
NM539	F$^-$, *supF*, *hsdR* (r$_K$$^-$, m$_K$$^+$), *lacY*, (P2)	Pro
NovaBlue	*endA1*, *hsdR17* (r$_{K12}$$^-$, m$_{K12}$$^+$) *supE44*, *thi-1*, *recA1*, *gyrA96*, *relA1*, *lac* [F', *proAB*, *lacI^{q}ZΔM15*::Tn10 (Tc^r)]	Nov
NovaBlue (DE3)	*endA1*, *hsdR17* (r$_{K12}$$^-$, m$_{K12}$$^+$) *supE44*, *thi-1*, *recA1*, *gyrA96*, *relA1*, *lac* [F', *proAB*, *lacI^{q}ZΔM15*::Tn10 (Tc^r)] (DE3)	Nov
P2392	F$^-$, *e14$^-$* (mcrA), *hsdR514*, *supE44*, *supF58*, *lacY1* or Δ(lacIZY)6, *galK2*, *galT22*, *metB1*, *trp55*, (P2 lysogen)	Str

PLK-A	F^-, $e14^-$ (mcrA), recA, lac, mcrB1, hsdR2 (r_K^-, m_K^+), supE44, galK2, galT22, metB1	Str
PLK-F'	$e14^-$ (mcrA), mcrB1, recA, lac, hsdR2 (r_K^-, m_K^+), supE44, galK2, galT22, metB1 [F', proAB, lacI^{q}ZΔM15, Tn10 (Tc^r)]	Str
PR700	F^-, Δ(gpt-proA)62, leu, supE44, ara14, galK2, lacY1, Δ(mcrC-mrr), rpsL20 (Str^r), xyl-5, mtl-1, RecA$^+$, Δ(malB), Δ(argF-lac)U169, Pro$^+$	Bio
PR745	F^-, Δ(gpt-proA)62, leu, supE44, ara14, galK2, lacY1, Δ(mcrC-mrr), rpsL20 (Str^r), xyl-5, mtl-1, RecA$^+$, lon::miniTn10 (Tc^r), Δ(malB), Δ(argF-lac)U169, Pro$^+$	Bio
RR1	F^-, mcrB, mrr, hsdS20 (r_B^-, m_B^-), supE44, ara-14, proA2, rpsL20 (Str^r), lacY1, galK2, xyl5, leu, mtl-1	BRL
SCS1	F^-, recA1, endA1, gyrA96, thi-1, hsdR17 (r_K^-, m_K^+), supE44, relA1	Str
SDM2	ϕ80dlacZΔM15, mcrA, Δ(mrr-hsdRMS-mcrBC), Δ(lac-proAB), Δ(recA 1398), deoR, rpsL, sri, thi [F', proAB$^+$, lacI^{q}ZΔM15]	USB
SRB	$e14^-$ (mcrA), Δ(mcrCB-hsdSMR-mrr)171, sbcC, recJ, uvrC, umuC::Tn5(Km^r), supE44, lac, gyrA96, relA1, thi-1, endA1 [F', proAB lacI^{q}ZΔM15]	Str
SRB(P2)	$e14^-$ (mcrA), Δ(mcrCB-hsdSMR-mrr)171, sbcC, recJ, uvrC, umuC::Tn5(Km^r), supE44, lac, gyrA96, relA1, thi-1, endA1 [F', proAB lacI^{q}ZΔM15], (P2 lysogen)	Str
SURE™	$e14^-$ (mcrA), Δ(mcrCB-hsdSMR-mrr)171, sbcC, recB, recJ, umuC::Tn5(Km^r), uvrC, supE44, lac, gyrA96, relA1, thi-1, endA1 [F', proAB, lacI^{q}ZΔM15, Tn10, (Tc^r)]	Str
TB1	F^-, ara, Δ(lac-proAB), rpsL (Str^r) [ϕ80dlacΔ(lacZ)M15], hsdR (r_K^-, m_K^+)	Bio
TOP10	F^-, mcrA, Δ(mrr-hsdRMS-mcrBC), ϕ80ΔlacZΔM15, ΔlacX74, deoR, recA1, araD139, Δ(ara, leu)7697, galU, galK, λ^-, rpsL, endA1, nupG	Inv
TOP10F'	F' [Tc^r], mcrA, Δ(mrr-hsdRMS-mcrBC), ϕ80Δlac-ΔM15, ΔlacX74, deoR, recA1, araD139, Δ(ara,leu)7697, galU, galK, λ^-, rpsL, endA1, nupG	Inv
UT5600	F^-, ara-14, leuB6, azi-6, lacY1, proC14, tsx-67, Δ(ompT-fepC)266, entA403, trpE38, rfbD1, rpsL109, xyl-5, mtl-1, thi-1	Bio
XL1-Blue	recA1, endA1, gyrA96, thi-1, hsdR17, supE44, relA1, lac [F', proAB, lacI^{q}ZΔM15, Tn10 (Tc^r)]	Str

Continued

135

Table 2. Genotypes of *E. coli* host strains, *continued*

Strain	Genotype	Supplier
XL1-Blue MRF'	Δ(mcrA)182, Δ(mcrCB-hsdSMR-mrr)172, endA1, supE44, thi-1, recA, gyrA96, relA1, lac, λ⁻, [F', proAB, lacIqZΔM15, Tn10 (Tcʳ)]	Str
XL1-Blue MR	F⁻, Δ(mcrA)182, Δ(mcrCB-hsdSMR-mrr)172, endA1, supE44, thi-1, recA, gyrA96, relA1, lac, λ⁻	Str
XS127/P3	[F', traD36, proAB⁺, lacIqZΔM15], gyrA, thi, rpoB331, argE, (rₖ⁺, mₖ⁺), supE44, Δ(lac-proAB) [P3: amber Apʳ, amber Tcʳ, Kmʳ]	Inv
Y1088	F', Δ(lacU169), supE, supF, hsdR (rₖ⁻, mₖ⁺), metB, trpR, tonA21, [proC::Tn5], (pMC9)	Inv, Clo, Str
Y1089	F', Δ(lacU169), proA⁺, Δ(lon), araD139, strA, hflA150, [chr::Tn10(Tcʳ)], (pMC9)	Inv, Str, Pro
Y1089r−	F', Δ(lacU169), proA⁺, Δ(lon), araD139, strA, hflA150, [chr::Tn10(Tcʳ)], (pMC9), mcrB	Str, Nov
Y1090	F', Δ(lacU169), proA⁺, Δ(lon), araD139, strA, supF, mcrA, [trpC22::Tn10(Tcʳ)], (pMC9), hsdR (rₖ⁻, mₖ⁺)	Inv, Str, Pro
Y1090r−	F', Δ(lacU169), proA⁺, Δ(lon), araD139, strA, supF, mcrA, [trpC22::Tn10(Tcʳ)],(pMC9), hsdR (rₖ⁻, mₖ⁺), mcrB	Clo, Str, Nov

Table 3. Lambda vector genotypes

Vector	Genotype
λ2001	λ.sbhIλ1°, b189 (polycloning site srlλ3°, ninL44, bio, polycloning site) KH54, chiC, srlλ4°, nin5, shndIIIλ6°, srlλ5°
λDASH	λ.sbhIλ1°, b189 (T3 promoter–polycloning site, srlλ3°, ninL44, bio, polycloning site–T7 promoter) KH54, chiC, srlλ4°, nin5, shndIIIλ6°, srlλ5°
λEMBL3/4	λ.sbhIλ1°, b189 (polycloning site, int29, ninL44, trpE, polycloning site) KH54, chiC, srlλ4°, nin5, srlλ5°
λFIX	λ.sbhIλ1°, b189 (T7 promoter–polycloning site, srlλ3°, ninL44, bio, polycloning site–T3 promoter) KH54, chiC, srlλ4°, nin5, shndIIIλ6°, srlλ5°
λgt10	λ.srlλ1°, b527, srlλ3°, imm434 (srl434⁺), srlλ4°, srlλ5°
λgt11	λlac5, ΔshndIIIλ2−3, srlλ3°, cIts857, srlλ4°, nin5, srlλ5°, Sam100

λgt18/19	λ*lac*5, Δ*shn*dIIIλ2−3, *srl*λ3°, Δ*ssl*Iλ1−2, c*l*ts857, *srl*λ4°, *nin*5, *srl*λ5°, Sam100
λgt20/21/22/23	λ*lac*5, Δ*sxl*λ1°, *chi*, Δ*shn*dIIIλ2−3, *srl*λ3°, Δ*ssl*Iλ1−2, c*l*ts857, *srl*λ4°, *nin*5, *srl*λ5°, Sam100
λgtWES.λb'	λWam403, Eam1100, *inv* (*srl*λ1−*srl*λ2), Δ(*srl*λ2−*srl*λ3), c*l*ts857, *srl*λ4°, *nin*5, *srl*λ5°, Sam100
λZAPII	λ*sbh*lλ1°, *chi*A131 (T, *amp*, ColE1, *ori*, *lacZ'*, T3 promoter−polycloning site−T7 promoter, I, *srl*λ3°, c*l*ts857, *srl*λ4°, *nin*5, *srl*λ5°, Sam100

Table 4. Key to genetic markers

Genotype/phenotype/marker	Description
(a) *E. coli* − related markers	
ara	Arabinose utilization
arg	Arginine biosynthesis
aro	Aromatic amino acid biosynthesis
asd	Aspartate semialdehyde biosynthesis
azi	Azide biosynthesis
bio	Biotin biosynthesis
chl	Chlorate resistance
cyc	Transport of D−alanine, D−serine and glycine
dam	DNA adenine methylase
dap	Diaminopimelate biosynthesis
dcm	DNA cytosine methylase
DE3	Carries integrated gene for T7 RNA polymerase
deo	Deoxyribose biosynthesis
dnaJ	One of several 'chaperonins' is inactive
dut	Deoxyuridinetriphosphatase
Continued	

137

Host Cell Information

Table 4. Key to genetic markers, *continued*

Genotype/phenotype/marker	Description
e14	An excisable prophage−like element carrying the *mcrA* gene
end	DNA−specific endonuclease I
env	Cell envelope
F	F plasmid
flb	Flagella synthesis
gal	Galactose utilization
gln	Glutamine biosynthesis and activation
gpt	Guanine–hypoxanthine phosphoribosyl transferase
gyr	DNA gyrase
hfl	High frequency of lysogeny by phage λ
his	Histidine biosynthesis
hsd	Host-specific restriction and/or modification. (r^+, m^-) indicates restriction enzyme present but corresponding methylase is absent
htp	Regulatory gene for proteins induced at high temperatures
ilv	Isoleucine and valine biosynthesis
lac	Lactose biosynthesis
lacZΔM15	Specific N-terminal deletion of β-galactosidase that permits α-complementation
lam	Phage receptor protein (maltose uptake)
leu	Leucine biosynthesis
lig	DNA ligase
lon	ATP-dependent protease
mal	Maltose utilization
mcr	Methylcytosine-specific restriction systems

met	Methionine biosynthesis
min	Formation of mini-cells containing no DNA
mrr	Cytosine or adenine methylation genes
mtl	Mannitol utilization
(Mu)	Mu prophage
mut	High mutation rate
omp	Outer membrane proteins
oms	Osmotic sensitivity
(P1)	Carries P1 prophage
(P2)	Carries P2 prophage
(P3)	Carries P3 plasmid
pgl	6-Phosphogluconolactonase
pho	Phosphate utilization
pLys	Contains pACYC184 (Cm^r) carrying the T7 lysozyme gene
pMC9	Plasmid encoding *lacI^q*
pnp	Polynucleotide phosphorylase
pro	Proline biosynthesis
pts	Phosphotransferase system
rbs	Ribose utilization
rec	General recombination and radiation repair
rel	Regulation of RNA synthesis
rfb	Cell wall biosynthesis
rha	Rhamnose utilization
rif	Rifampicin sensitivity or resistance
rna	Ribonuclease I
rpoH	Heat-shock transcription factor
Continued	

139

Host Cell Information

Table 4. Key to genetic markers, *continued*

Genotype/phenotype/marker	Description
rpoB	RNA polymerase
rps	Small ribosomal protein
rrn	rRNA
sbc	Suppressor of *recBC*
srl	Sorbitol utilization
sup	Suppressor
thi	Thiamine requirement
thr	Threonine biosynthesis
thy	Thymidylate synthase
Tn5	Transposon encoding kanamycin resistance
Tn10	Transposon encoding tetracycline resistance
Tn1000	Transposon containing the $\gamma\delta$ element
ton	Resistance to T1 phage
tra	Transfer genes (typically of F plasmids)
trp	Tryptophan biosynthesis
tsp	Periplasmic protease
tsx	Resistance to T6 phage and to colicin K
tyr	Tyrosine biosynthesis and activation
umu	Sensitivity to UV
ung	Uracil–DNA glycosylase
uvr	Repair of UV radiation damage to DNA
xyl	Xylose utilization
(ϕ80)	Carries the lambdoid prophage ϕ80

λimm^{434}(Km)	Lysogen conferring kanamycin resistance and immunity to infection by phage 434 immunity group

(b) Lambda phage-related markers

Aam	Amber mutation in gene A
*b*2, *b*189, *b*527, *b*1007	Deletions preventing the phage entering the lysogenic cycle
Bam	Amber mutation in gene B
bio, *bio*252, *bio*256	Substitutions from the *bio* region of *E. coli*
chiA131, *chi*C	Directional recombination sites
cl*ts*857	Temperature-sensitive mutation in cl allowing induction by heat-shock
*cos*2	Defective *cos* sites
Dam	Results in intracellular accumulation of λ structural proteins
Eam	Results in intracellular accumulation of λ structural proteins
exo, *bet*, *red*3, *gam*	Mutations result in Spi$^-$ phenotype
*imm*434	Substitution from phage ϕ434
*int*29	Substitution from phage ϕ29
int	Amber mutation in *int*
KH54	Deletion that prevents lysogeny
*lac*5, *lac*UV5, *lacZ*, *lacZ'*	Substitutions from the *lac* region of *E. coli*
nin5, *nin*L44	Allows delayed early transcription independent of the N gene product
QSR80	Substitution from phage ϕ80
s....λ	Designates deletion between restriction endonuclease sites
*Sam*100	Results in intracellular accumulation of phage particles
spi$^-$	Refers to *red*$^-$ *gam*$^-$ mutant derivatives of λ
trpE	Substitutions from the *trp* region of *E. coli*
Wam	Amber mutation in gene *W*
*WL*113	Deletions that removes *kil*, clll, *Ea*10 and *ral*

141 *Host Cell Information*

Chapter 5 **SUPPLIERS**

Ame **Amersham International plc,** Amersham Place, Little Chalfont, Buckinghamshire HP7 9NA, UK.
Tel: (01494) 544000.
Fax: (01494) 542266.
2636 South Clearbrook Drive, Arlington Heights, IL 60005, USA.
Tel: (708) 5936300.
Fax: (708) 5938010.

Apl **Appligene Inc.,** Pinetree Centre Durham Road, Birtley, Chester-le-Street, Durham DH3 2TD, UK.
Tel: (0191) 4920022.
Fax: (0191) 4920617.
1177-C Quarry Lane, Pleasanton, CA 94566, USA.
Tel: (510) 4622232.
Fax: (510) 4626247.

BRL **Life Technologies Ltd,** PO Box 35, Trident House, Renfrew Road, Paisley PA3 4EF, UK.
Tel: (0141) 8146100.
Fax: (0141) 8871167.

GIBCO Laboratories, Life Technologies Inc., 3175 Staley Road, Grand Island, NY 14072, USA.
Tel: (716) 7730700.

Bio **New England Biolabs,** 67 Knowl Piece, Wilbury Way, Hitchin, Hertfordshire SG4 OTY, UK.
Tel: (01462) 420616.
Fax: (01462) 421057.
32 Tozer Road, Beverly, MA 01915-5599, USA.
Tel: (508) 9275054.
Fax: (508) 9211350.

Boe **Boehringer Mannheim,** Bell Lane, Lewes, East Sussex BN7 1LG, UK.
Tel: (01273) 480444.
Fax: (01273) 480266.
P.O.Box 50414, Indianapolis, IN 46250, USA.
Tel: (800) 2621640.
Fax: (317) 5762754.

Clo **Clontech & Pharmingen Distributors,** Cambridge Biosciences, 25 Signet Court, Newmarket Road, Cambridge CB5 8LA, UK.
Tel: (01223) 316855.
Fax: (01223) 60732.
Clontech Laboratories Inc., 4030 Fabian Way, Palo Alto, CA 94303-4607, USA.
Tel: (415) 4248222.
Fax: (415) 4241352.

Inv (Invitrogen Distributors in UK) **R&D Systems Europe Ltd,** 4–10 The Quadrant, Barton Lane, Abingdon OX14 3YS, UK.
Tel: (01235) 531074.
Fax: (01235) 533420.
Invitrogen Corporation, 3985 B Sorrento Valley Blvd, San Diego, CA 92121, USA.
Tel: (800) 9556288.
Fax: (619) 5976201.

NBL **NBL Gene Sciences Ltd,** South Nelson Industrial Estate, Cramlington, Northumberland NE23 9HL, UK.
Tel: (01670) 733015.
Fax: (01670) 730454.

NBS **New Brunswick Scientific Ltd,** Edison House, 163 Dixons Hill Road, North Mymms, Hatfield AL9 7JE, UK.
Tel: (01707) 275733/75707.
Fax: (01707) 267859.

Nov **Novagen Inc.** (NBL is UK distributor for Novagen's products), 597 Science Drive, Madison WI 53711, USA.
Tel: (800) 5267319.

Pro **Promega Ltd,** Delta House, Enterprise Road, Chilworth Research Centre, Southampton SO1 7NS, UK.
Tel: (01703) 760225.
Fax: (01703) 767014.
2800 Woods Hollow Road, Madison, WI 53711-5399, USA.
Tel: (608) 2744330,
　　 (800) 3569526.
Fax: (608) 2736967.

Sig **Sigma Chemical Company Ltd,** Fancy Road, Poole, Dorset BH17 7NH, UK.
Tel: (0800) 373731.
Fax: (0800) 378785.
P O Box 14508, St Louis, MO 63178, USA.
Tel: (800) 3253010.
Fax: (800) 3255052.

Par The Pharmingen distributor in the UK is Clontech (Clo).
Pharmingen, 11555 Sorrento Valley Road, San Diego, CA 92121, USA.
Tel: (619) 7925730.
Fax: (619) 7925238.

Pha **Pharmacia Biotech Ltd,** Davy Avenue, Knowlhill, Milton Keynes MK5 8PH, UK.
Tel: (01908) 661101.
Fax: (01908) 690091.
800 Centennial Avenue, Piscataway, NJ 08854, USA.
Tel: (800) 5263593.
Fax: (908) 4570557.

Str **Stratagene Ltd,** Cambridge Innovation Centre, Cambridge Science Park, Milton Road, Cambridge CB4 4GF, UK.
Tel: (01223) 420955.
Fax: (01223) 420234.
11099 North Torrey Pines Road, La Jolla, CA 92037, USA.
Tel: (619) 5355400.
Fax: (619) 5350045.

USB The USB distributor in the UK is Amersham Life Science (Ame).
USB Corporation, PO Box 22400, Cleveland, OH 44122, USA.
Tel: (216) 7655000.
Fax: (216) 4645075.

REFERENCES

1. Lovett, M.A. *et al.* (1974) *Proc. Natl Acad. Sci. USA* **71**, 3854.
2. So, M. *et al.* (1975) *Molec. Gen. Gene*t. **142**, 239.
3. Anon. (1993) *Boehringer Mannheim Biochemica Catalogue. Biochemicals for Molecular Biology*. Boehringer-Mannheim, Germany.
4. Messing, J. and Vieira, J. (1982) *Gene* **19**, 269.
5. Messing, J. *et al.* (1981) *Nucl. Acids Res.* **9**, 309.
6. Norrander, J. *et al.* (1983) *Gene* **26**, 101.
7. Chang, A.C.Y. and Cohen, S.N. (1978) *J. Bacteriol.* **134**, 1141.
8. Rose, R.E. (1988) *Nucl. Acids Res.* **16**, 356.
9. Twigg, A.J. and Sherratt, D. (1980) *Nature* **283**, 216.
10. Bolivar, F. *et al.* (1977) *Gene* **2**, 95.
11. Sutcliffe, J.G. (1979) *Cold Spring Harbor Symp. Quant. Biol.* **43**, 77.
12. Soberon, X. *et al.* (1980) *Gene* **9**, 287.
13. Balbas, P. *et al.* (1986) *Gene* **50**, 3.
14. Anon. (1994/1995) *Molecular Biology Reagents Catalog*. United States Biochemical Corp., Cleveland, OH.
15. Dente, L. *et al.* (1983) *Nucl. Acids Res.* **11**, 1145.
32. Karn, J. *et al.* (1983) *Gene* **32**, 217.
33. Swaroop, A. and Weissman, S.M. (1988) *Nucl. Acids Res.* **16**, 8739.
34. Elgin, E. *et al.* (1991) *Strategies* **4**, 6.
35. Murphy, A.J.M. *et al.* (1992) *Methods: a Companion to Methods in Enzymology* **4**, 111.
36. Swirski, R.A. *et al.* (1992) *Methods: a Companion to Methods in Enzymology* **4**, 133.
37. Elgin, E. *et al.* (1991) *Strategies* **4**, 8.
38. Frischauf, A.M. *et al.* (1983) *J. Mol. Biol.* **170**, 827.
39. Zabarousky, E.R. and Allikmets, R.L. (1986) *Gene* **42**, 119.
40. Huynh, T.V. *et al.* (1985) in *DNA Cloning: a Practical Approach* (D.M. Glover, ed.), Vol. I, p. 49. IRL Press, Oxford.
41. Murray, N.E. *et al.* (1977) *Molec. Gen. Genet.* **150**, 53.
42. Young, R.A. and Davis, R.W. (1983) *Proc. Natl Acad. Sci. USA* **80**, 1194.
43. Sambrook, J., Fritsch, E.F. and Maniatis, T. (1989) *Molecular Cloning: a Laboratory Manual*. Cold Spring Harbor Laboratory Press, Cold Spring Harbor, NY.

16. Dente, L. *et al.* (1985) in *DNA Cloning: a Practical Approach* (D.M. Glover, ed.), Vol. I, p. 101. IRL Press, Oxford.

17. Anon. (1993/1994) *Biological Research Products Catalog*. Promega Corp., Madison, WI.

18. Bolivar, F. *et al.* (1977) *Gene* **2,** 75.

19. Anon. (1994) *Molecular Biology Catalog*. Amersham Life Sciences, Buckinghamshire, UK.

20. Marchuk, D. *et al.* (1991) *Nucl. Acids Res.* **19,** 1154.

21. Anon. (1992) *Stratagene Catalog for Molecular Biology*. Stratagene, La Jolla, CA.

22. Anon. (1994) *Novagen Catalog*. Novagen, Madison, USA.

23. Brosius, J. (1989) *DNA* **8,** 759.

24. Anon. (1993) *GIBCO-BRL Catalog for Molecular Biology*. GIBCO-BRL, Gaithersburg, MD.

25. Melton, D.A. *et al.* (1984) *Nucl. Acids Res.* **12,** 7035.

26. Hoffman, L. and Donaldson, D. (1988) *Gene* **67,** 137.

27. Krieg, P. and Melton, D. (1987) *Meth. Enzymol.* **155,** 397.

28. Anon. (1992/1993) *Molecular Biologicals Catalog*. Pharmacia LKB, Sweden.

29. Mead, D.A. *et al.* (1986) *Protein Engin.* **1,** 67.

30. Rokeach, L.A. *et al.* (1988) *Proc. Natl Acad. Sci. USA* **85,** 4832.

31. Ruther, U. *et al.* (1981) *Nucl. Acids Res.* **9,** 4087.

44. Anon. (1993/1994) *Tools for the Molecular Biologist*. Clontech, Palo Alto, CA.

45. Palazzolo, M.J. *et al.* (1990) *Gene* **88,** 25.

46. Anon. (1994) *Invitrogen Molecular Biology Solutions Catalog*. San Diego, CA.

47. Short, J. M. *et al.* (1988) *Nucl. Acids Res.* **16,** 7583.

48. Sorge, J. (1988) *Strategies* **1,** 3.

49. Hohn, B. and Collins, J. (1980) *Gene* **11,** 291.

50. Ish-Horowicz, D. and Burke, J.F. (1981) *Nucl. Acids Res.* **9,** 2989.

51. Lau, Y.F. and Kan, Y.W. (1983) *Proc. Natl Acad. Sci. USA* **80,** 5225.

52. Wahl, G.M. *et al.* (1987) *Proc. Natl Acad. Sci. USA* **84,** 2160.

53. Germino, J. and Bastia, D. (1984) *Proc. Natl Acad. Sci. USA* **81,** 4692.

54. Markmeyer, P. *et al.* (1990) *Gene* **93,** 129.

55. Scholtissek, S. and Grosse, F. (1988) *Gene* **62,** 55.

56. Alting-Meese, M.A. and Short, J.M. (1989) *Nucl. Acids Res.* **17,** 9494.

57. de Boer, H.A. *et al.* (1983) *Proc. Natl Acad. Sci. USA* **78,** 21.

58. Hallewell, R.A. and Emtage, S. (1980) *Gene* **9,** 27.

59. Tacon, W. *et al.* (1980) *Molec. Gen. Genet.* **177,** 427.

60. Okayama, H. and Berg, P. (1983) *Mol. Cell. Biol.* **3,** 280.

61. Elroy-Stein, O. *et al.* (1989) *Proc. Natl Acad. Sci. USA* **86,** 6126.
62. Parks, C.L. *et al.* (1986) *J. Virol.* **60,** 376.
63. Babineau, D. *et al.* (1985) *J. Biol. Chem.* **260,** 12313.
64. Groger, R. *et al.* (1989) *Gene* **81,** 285.
65. Margolskee, R. *et al.* (1988) *Mol. Cell. Biol.* **8,** 2837.
66. Richards, R. *et al.* (1984) *Cell* **37,** 263.
67. Yates, J. *et al.* (1985) *Nature* **313,** 812.
68. Studier, F.W. *et al.* (1990) *Meth. Enzymol.* **185,** 60.
69. Lusky, M. and Botchan, M. (1981) *Nature* **293,** 79.
70. Templeton, D. and Eckhart, W. (1984) *Mol. Cell. Biol.* **4,** 817.
71. Haymerle, H. *et al.* (1986) *Nucl. Acids Res.* **14,** 8615.
72. Stanley, K.K. and Luzio, J.P. (1984) *EMBO J.* **3,** 1429.
73. Nilsson, B. *et al.* (1991) *Meth. Enzymol.* **198,** 3.
74. Eaton, D. *et al.* (1986) *Biochemistry* **25,** 505.
75. Nagai, K. and Thoegersen, C. (1986) *Nature* **309,** 810.
76. Brosius, J. and Holy, A. (1984) *Proc. Natl Acad. Sci. USA* **81,** 6929.
77. Rodriguez, R.L. and Denhardt, D.T. (eds) (1988) *Vector: a Survey of Molecular Cloning Vectors and Their Uses*, p. 213.
78. Shimizu, Y. *et al.* (1988) *Gene* **65,** 141.
79. Guan, C. *et al.* (1988) *Gene* **67,** 21.
80. Maina, C.V. *et al.* (1988) *Gene* **74,** 365.
100. Luckow, V. and Summers, M. (1988) *Bio/Technology* **6,** 47.
101. Anon. (1992/1993) *Pharmingen Catalog for Molecular Biology*. San Diego, CA.
102. Weyer, U. *et al.* (1991) *J. Gen. Virol.* **72,** 2967.
103. Alam, J. (1990) *Anal. Biochem.* **188,** 245.
104. MacGregor, G.R. and Caskey, C.T. (1989) *Nucl. Acids Res.* **17,** 2365.
105. Norton, P.A. and Coffin, J.M. (1985) *Mol. Cell. Biol.* **5,** 281.
106. Jefferson, R.A. (1989) *Nature* **342,** 857.
107. Jefferson, R.A. *et al.* (1987) *EMBO J.* **6,** 3901.
108. Jefferson, R.A. (1987) *Plant Mol. Biol. Reporter* **5,** 387.
109. Lancaster, W.D. (1981) *Virology* **108,** 251.
110. Lowy, D.R. *et al.* (1980) *Nature* **287,** 72.
111. Fromm, M.E. *et al.* (1985) *Proc. Natl Acad. Sci. USA* **82,** 5824.
112. Fromm, M.E. *et al.* (1986) *Nature* **319,** 791.
113. Hall, C.V. *et al.* (1983) *J. Mol. Appl. Gen.* **2,** 101.
114. Herbomel, P. *et al.* (1984) *Cell* **39,** 653.
115. Oka, A. *et al.* (1991) *J. Mol. Biol.* **147,** 217.
116. Clark, J.M. (1988) *Nucl. Acids Res.* **16,** 9677.
117. Gahm, S.J. *et al.* (1991) *Proc. Natl Acad. Sci. USA* **88,** 10267.
118. Jarolim, P. *et al.* (1991) *Proc. Natl Acad. Sci. USA* **88,** 11022.
119. Anon. (1994/1995) *Molecular Biology Catalog*. Appligene, Pleasanton, CA.

81. Zagursky, R.J. and Berman, M.L. (1984) *Gene* **27,** 183.

82. Lee, F. *et al.* (1981) *Nature* **294,** 228.

83. Narayanan, R. *et al.* (1992) *Oncogene* **7,** 553.

84. Casadaban, M.J. *et al.* (1983) *Meth. Enzymol.* **100,** 293.

85. Shapira, S.K. *et al.* (1983) *Gene* **25,** 71.

86. Hasan, N. and Szybalski, W. (1987) *Gene* **56,** 145.

87. Podhajsku, A.J. *et al.* (1985) *Gene* **40,** 163.

88. Nilsson, B. *et al.* (1985) *EMBO J.* **4,** 1075.

89. Green, S. *et al.* (1988) *Nucl. Acids Res.* **16,** 369.

90. Sprague, J. *et al.* (1983) *J. Virol.* **45,** 773.

91. Amann, E. *et al.* (1988) *Gene* **69,** 301.

92. Helfman, D.M. *et al.* (1983) *Proc. Natl Acad. Sci. USA* **80,** 31.

93. Vieira, J. and Messing, J. (1982) *Gene* **19,** 259.

94. Yanisch-Perron, C. *et al.* (1985) *Gene* **33,** 103.

95. Pruitt, S. (1988) *Gene* **66,** 121.

96. Boulter, C.A. and Wagner, E.F. (1987) *Nucl. Acids Res.* **17,** 7194.

97. Keller, G. *et al.* (1985) *Nature* **318,** 149.

98. O'Reilly, D.R. *et al.* (1992) *Baculovirus Expression Vectors: a Laboratory Manual.* W.H. Freeman, New York.

99. Anon. (1992) *The Digest* **5,** 2:2.

120. Brau, B. *et al* (1984) *Mol. Gen. Genet.* **193,** 179.

121. Gritz, L. and Davies, J. (1983) *Gene* **25,** 179.

122. Kuhstoss, S. and Naharaja Rao, R. (1984) *Gene* **26,** 295.

123. McKnight, S.L. and Gavis, E.R. (1980) *Nucl. Acids Res.* **8,** 5931.

124. Brosius, J. (1984) *Gene* **27,** 151.

125. Thomas, K.R. and Capecchi, M.R. (1987) *Cell* **51,** 503.

126. Colbere-Garapin, F. *et al.* (1981) *J. Mol. Biol.* **150,** 1.

127. Southern, P.J. and Berg, P.J. (1982) *J. Mol. Appl. Gene*t. **1,** 327.

128. Rodriguez, J.F. *et al.* (1988) *Proc. Natl Acad. Sci. USA* **85,** 1667.

129. Gryczan, T.J. *et al.* (1978) *J. Bacteriol.* **134,** 318.

130. Albertson, H.M. *et al.* (1990) *Proc. Natl Acad. Sci. USA.* **87,** 4256.

131. Burke, D.T. *et al.* (1987) *Science* **236,** 806.

132. Weilguny, D. *et al.* (1991) *Gene* **99,** 47.

133. Peden, K.W.C. (1983) *Gene* **22,** 277.

134. Roberts, R.J. (1987) *Nucl. Acids Res.* **15** (Suppl), r189–r217.

135. Rose, M. *et al.* (1984) *Gene* **29,** 113.

APPENDIX A

pET vector classification

Vector	Selection	Promoter	Cloning sites for N-terminal fusion	Cleavage site	Optional C-terminal fusion
(a) Transcription vectors					
For expression from translation initiation signals within a cloned insert					
pET-23(+)	Ap	T7	*Bam*HI, *Eco*RI, *Sac*I, *Sal*I, *Hinc*II, *Hind*III, *Eag*I, *Not*I, *Xho*I		His·Tag®
pET-21(+)	Ap	T7*lac*	*Bam*HI, *Eco*RI, *Sac*I, *Sal*I, *Hind*III, *Eag*I, *Not*I, *Xho*I, *Ava*I		His·Tag
pET-24(+)	Km	T7*lac*	*Bam*HI, *Eco*RI, *Sac*I, *Sal*I, *Hind*III, *Eag*I, *Not*I, *Xho*I		His·Tag
(b) Translation vectors					
1. Vectors that allow N-terminal fusion to small (11 aa) T7·Tag® sequence for ability to follow expression with T7·Tag antibody					
pET-3a–d	Ap	T7	*Bam*HI		No
pET-5a–c	Ap	T7	*Bam*HI, *Eco*RI		No
pET-9a–d	Km	T7	*Bam*HI		No
pET-17b	Ap	T7	*Hind*III, *Kpn*I, *Sac*I, *Bam*HI, *Spe*I, *Bst*XI[1], *Eco*RI, *Eco*RV, *Not*I, *Xho*I		No
pET-23a–d(+)	Ap	T7	*Bam*HI, *Eco*RI, *Sac*I, *Sal*1, *Hinc*II, *Hind*III, *Eag*I, *Not*I, *Xho*I		His·Tag
pET-11a–d	Ap	T7*lac*	*Bam*HI		No
pET-21a–d(+)	Ap	T7*lac*	*Bam*HI, *Eco*RI, *Sac*I, *Sal*I, *Hind*III, *Eag*I, *Not*I, *Xho*I, *Ava*I		His·Tag
pET-24a–d(+)	Km	T7*lac*	*Bam*HI, *Eco*RI, *Sac*I, *Sal*I, *Hind*III, *Eag*I, *Not*I, *Xho*I		His·Tag

2. Vectors that allow N-terminal fusion to large (260 aa) T7·Tag sequence for stabilization of small target proteins/peptides

pET-3xa–c	Ap	T7	BamHI	No
pET-17xb	Ap	T7	SacII, HindIII, KpnI, SacI, BamHI, SpeI, BstXI[1], EcoRI, EcoRV, NotI, XhoI	No
pTOPE™-1b(+)[2]	Ap	T7	SacII, HindIII, SacI, BamHI, SpeI, BstXI[1], EcoRI, PstI, EcoRV, NotI, XhoI, AvaI	No

3. Vectors that allow fusion to N-terminal signal sequence for potential periplasmic localization

pET-12a–c	Ap	T7	BamHI	No
pET-20b(+)	Ap	T7	NcoI, EcoRV, BamHI, EcoRI, SacI, SalI, HincII, HindIII, EagI, NotI, XhoI	His·Tag
pET-22b(+)	Ap	T7lac	MscI[3], NcoI, BamHI, EcoRI, SacI, SalI, HindIII, EagI, NotI, XhoI, AvaI	His·Tag
pET-26b(+)	Km	T7lac	MscI[3], NcoI, BamHI, EcoRI, SacI, SalI, HindIII, EagI, NotI, XhoI	His·Tag
pET-25b(+)	Ap	T7lac	MscI[3], NcoI, BamHI, EcoRI, SacI, SalI, HindIII, EagI, NotI, XhoI, NheI	His·Tag + HSV·Tag™
pET-27b(+)	Km	T7lac	MscI[3], NcoI, BamHI, EcoRI, SacI, SalI, HindIII, EagI, NotI, XhoI, NheI	His·Tag + HSV·Tag

4. Vectors that allow N-terminal fusion to cleavable His·Tag sequence for rapid affinity purification

pET-14b	Ap	T7	NdeI, XhoI, BamHI	Thrombin	No
pET-15b	Ap	T7lac	NdeI, XhoI, BamHI	Thrombin	No
pET-16b	Ap	T7lac	NdeI, XhoI, BamHI	Factor Xa	No
pET-19b	Ap	T7lac	NdeI, XhoI, BamHI	Enterokinase	No

Continued

151

pET vector classification, *continued*

Vector	Selection	Promoter	Cloning sites for N-terminal fusion	Cleavage site	Optional C-terminal fusion
5. Vectors that allow N-terminal fusion to cleavable His·Tag and non-cleavable T7·Tag sequence					
pET-28a–c(+)	Km	T7*lac*	For His·Tag only: *Nde*I, *Nhe*I	Thrombin	His·Tag
			For His·Tag and T7·Tag: *Bam*HI, *Eco*RI, *Sac*I, *Sal*I, *Hind*III, *Eag*I, *Not*I, *Xho*I		
6. Vectors that allow N-terminal fusion to cleavable S·Tag™ sequence for rapid assay and affinity purification					
pET-29a–c(+)[2]	Km	T7*lac*	*Nco*I, *Eco*RV, *Bam*HI, *Eco*RI, *Sac*I, *Sal*I, *Hind*III, *Eag*I, *Not*I, *Xho*I	Thrombin	His·Tag

[1]pET-17b, pET-17xb and pTOPE-1b(+) contain dual *Bst*XI sites designed for efficient cloning using nonpalindromic linkers (Seed (1987) *Nature* **329,** 840).

[2]pTOPE-1b(+) and pET-29c(+) are available as T-Vectors, which are prepared for cloning inserts having single 3′ dA overhangs (e.g. PCR products and DNA fragments treated with the Single dA™ Tailing Kit). The T-cloning site is *Eco*RV.

[3]When cloning into the *Msc*I site, the insert should start with dC to recreate the *Ala* residue and allow posttranslational processing by signal peptidase.

All vectors also contain *Nde*I or *Nco*I cloning sites for nonfusion subcloning.

Reproduced with permission from the Novagen catalog, 1994.

APPENDIX B

Sequencing primers

Primer	Sequence
Bac1	5' ACCATCTCGCAAATAAATAAG 3'
Bac2	5' ACAACGCACAGAATCTAGCG 3'
CITE	5' GGGGACGTGGTTTTCCTTTG 3'
DR2 F	5' CTGGTAAGTTTAGTCTTTTTGTC 3'
DR2 R	5' CAGTGCCAAGCTTGCATGCCT 3'
EBV R	5' GTGGTTTGTCCAAACTCATC 3'
GL1	5' TGTATCTTATGGTACTGTAACTG 3'
GL2	5' CTTTATGTTTTTGGCGTCTTCCA 3'
gt10 F	5' CTTTTGAGCAAGTTCAGCCTGGTTAAG 3'
gt10 R	5' GAGGTGGCTTATGAGTATTTCTTCCAGGGTA 3'
gt11 F	5' GGTGGCGACGACTCCTGGAGCCCG 3'
gt11 R	5' TTGACACCAGACCAACTGGTAATG 3'
GUS	5' TCACGGGTTGGGGTTTCTAC 3'
KS	5' CGAGGTCGACGGTATCG 3'
M13 F	5' GTAAAACGACGGCCAGT 3'
M13 R	5' AACAGCTATGACCATG 3'
malE	5' GGTCGTCAGACTGTCGATGAAGCC 3'
p10	5' GTATATTAATTAAAATACTATACTG 3'
pAX1	5' CCTGGTCTTGCTGGCCAACAT 3'
PAX2	5' CCCGGCGGCAACCGAGCGTTCT 3'

Primer	Sequence
pCDM8 R	5' TAAGGTTCCTTCACAAAG 3'
pH F	5' AAATGATAACCATCTCGC 3'
pH R	5' GTCCAAGTTTCCCTG 3'
pMEX F	5' CGGCTCGTAATAATGTGTGG 3'
PMEX R	5' TCTTCTCTCATCCGCC 3'
pREP F	5' GCTCGATACAATAAACGCC 3'
pRSET R	5' TAGTTATTGCTCAGCGGTGG 3'
Pst ccw	5' AACGACGAGCGTGAC 3'
Pst cw	5' GCTAGAGTAAGTAGTT 3'
pTrc His F	5' GAGGTATATATTAATGTATCG 3'
pXPRS +	5' GCCTGTACGGAAGTGTTA 3'
pXPRS −	5' GCTGGTTCTTTCCGCCTCA 3'
Sal ccw	5' AGTCATGCCCCGCGC 3'
Sal cw	5' ATGCAGGAGTCGCAT 3'
SK	5' TCTAGAACTAGTGGATC 3'
SP6	5' GATTTAGGTGACACTATAG 3'
T3	5' ATTAACCCTCACTAAAG 3'
T7	5' AATACGACTCACTATAG 3'
T7 gene 10	5' TGAGGTTGTAGAAGTTCCG 3'

cw, clockwise; ccw, counterclockwise.

INDEX

Index